INTERNATIONAL PERSPECTIVES ON INFORMATION SYSTEMS

International Perspectives on Information Systems

A Social and Organisational Dimension

Edited by
Savvas Katsikides
University of Cyprus

Graham Orange
Leeds Metropolitan University

Ashgate

Aldershot • Brookfield USA • Singapore • Sydney

303.4833
I612

Published by
Ashgate Publishing Limited
Gower House
Croft Road
Aldershot
Hants GU11 3HR
England

Ashgate Publishing Company
Old Post Road
Brookfield
Vermont 05036
USA

British Library Cataloguing in Publication Data
International perspectives on information systems : a
    social and organisational dimension
    1. Information technology - Social aspects 2. Business
    enterprises - Data processing
    I. Katsikides, Savvas II. Orange, Graham
    303.4'834

Library of Congress Catalog Card Number: 98-07146

ISBN 1 84014 508 0

Printed and bound by Athenaeum Press, Ltd.,
Gateshead, Tyne & Wear.

Contents

PART III: USER PARTICIPATION

List of Figures and Tables

Foreword

The book addresses the increasingly important social and organisational issues of information systems development both at the operational level and within a wider social context. It is not our intention that this book be a highly specialised text covering a narrow subject area. Rather it is intended that we give a feel for current research within a wider scope. Unlike many texts which are a collection of 'definitive', perhaps even 'tired', works this text contains new papers reflecting the current state of research in diverse, yet related, fields of study.

We decided against producing a highly specialised text because that would have a very limited appeal. We also wanted to bring together a number of complementary subject areas. We believe that much of today's current research, and indeed many courses, are interdisciplinary and we wanted to reflect this within this book. In fact the very nature of research is such that the more perceptive are able to build on findings from a wide spectrum of related interests.

Socio-technical thinking was popularised in the seventies and is still a contemporary issue. We saw the emergence of such terms as democratic technology, psychological technology, technology as constitution of the world, or more precisely, sociology of technology. The effects of new technology, its reproduction and its adaptation in the production process (of both products and services) affects various sectors of society. One major effect is the so called societal deregulation of labour relations impacting on the sphere of employers' activities. The social foundations of technology seems to be a newly established interest area of social and technical sciences. This volume brings together new research and studies which are concerned with the design, production and implementation of information technology within the context of the emerging social dimension.

The book is aimed at supporting information systems courses incorporating a behavioural element or sociology courses with an information systems content. Increasingly, information systems curricula developers are recognising that technology implementation cannot be addressed in isolation from social and organisational issues. Course design frequently includes modules that cover such topics as strategic planning, organisation culture, and organisation behaviour in order that information systems may be studied within the wider organisational context.

In a book of this nature we cannot hope to provide a definitive and complete text but wish to provide a flavour of current research into information systems with a social and organisational perspective in the hope that the reader will find the articles of interest and of relevance to their own area of study. To some it will support their work, to others we hope that it will inspire them to further research.

The main objective of this book is to publish work which reflects the environmental issues surrounding the development of information systems and the implementation of information technology. There are three main subdivisions of this book. Part I - Sociotechnical Perspective, Part II - Organisational Perspective and Part III - User Participation.

Each of these themes are important because traditionally we have seen systems and technology being imposed with scant regard to the environment in which they are being placed. Within the book we examine issues at various levels, from consideration of individual systems through to application of technology within society. Each of which is of importance to systems applications design and implementation.

We have received contributions from a number of people who are respected within their field of research. The contributors are located throughout Europe and the United States giving an international perspective. We start the book with articles looking at the societal impact of information systems and technology and then progress to looking at social issues within the organisation moving through to articles which address the organisational and methodological issues. Thus addressing a wide spectrum of systems and technology issues providing, we hope, a well rounded text.

Part I - Sociotechnical Perspective

In this section sociological considerations of technology implementation are discussed. We begin at the macro level by examining technology within society. This is important because macro level research is often the foundation of study at the organisation level. Principles of technology implementation within society are discussed in the first two articles by Josef Hochgerner and Savvas Katsikides. Hochgerner suggests that rather than look at two separate spheres, one social and the other technical, one should look at technology in society within the context of sociotechnical systems. Savvas Katsikides further develops this concept of sociotechnical systems by examining change requirements by organisations for technological implementation with reference to information technology. Jane Young looks at the implementation of information technology within the organisation suggesting that there is a cultural gap between the systems developers and senior management and users leading to mistrust and to the development of systems which do not satisfy the requirements of the business. Using a case study she looks at how this gap may be bridged.

Part II - Organisational Perspective

In this section articles address organisational considerations in addition to social issues. The article by Rob Kling suggests that computer science graduates are inadequately equipped for developing organisational systems and argues the case for the teaching of organisational and behavioural analysis on computer science (and similar) courses to increase the usability of computer systems. Harold Salzman also stresses the importance of the need for an understanding of organisations, and hence the social contextualisation of software design such that are flexible and acceptable to users. Margit Pohl examines the need for economic and organisational change in order that information technology may be implemented successfully. Graham Orange et al and David Hobbs et al look at the development of information systems within the context of organisational objectives and as part of an information systems strategy. Both articles also address culture and people issues, Orange through the mechanism of Geographic Information Systems and Hobbs through Multimedia systems. Werner Beuschel stresses the importance of understanding the organisational context in which systems are to be integrated and demonstrates this through case studies of expert systems implementation.

Part III - User Participation

User participative systems development approaches are considered important for ensuring that computer based systems meet organisational and social requirements. The remaining articles follow this theme. Hanseth and Monteiro examine information infrastructure and inter-organisational collaboration whilst Edeltraud Egger maps the transition from the user as a provider of information to the user taking a proactive role in systems development. Chris Dixon et al looks at Computer Supported Co-operative Work (CSCW) from the point of view of the developers (both technical and users). Then at prototyping as a development approach and discuss a particular tool for systems development support. Johannes Gärtner et al consider CSCW and group facilitation as a means of improving the performance of workgroups. By integrating CSCW and facilitation approaches the claim that the efficiency of workgroups is made more efficient.

The Editors

Savvas Katsikides is Associate Professor for Sociology at the University of Cyprus, Department for Social and Political Sciences in Nicosia, Cyprus. He studied Social and Economic Sciences at the Johannes Kepler University in Linz, Austria. He was Visiting Professor at the Leeds Metropolitan University, UK and at the Central Connecticut State University, New England, USA. His areas of interest are Sociology of Technology, Sociology of Work and the social inter-disciplinary issues of information technology.

Graham Orange is a Principal Lecturer at Leeds Metropolitan University. Prior to joining the University Graham spent many years as a systems developer, business systems analyst and systems consultant, specialising in information systems development methodologies and information strategy planning. Since joining the University Graham has continued offering systems consultancy and has supported this with research into cultural and behavioural issues of information systems development.

PART I
SOCIOTECHNICAL
PERSPECTIVE

1 Organising Technology: Prerequisites of Technology as an Agent of Social Change

JOSEF HOCHGERNER
CENTRE FOR SOCIAL INNOVATION,
VIENNA, AUSTRIA

Abstract

In a time when technical artefacts are of obviously great societal importance, it is a much discussed problem as to whether technological development is following a determined and dominant path, and, whether or not in the process of technical and social change comparable models are perceptible (cf. Katsikides/S., M. Campbell/J. Hochgerner/S., 1994). In such investigations it is not a question of comparing two separate spheres (one social, the other technical). Instead, the approach to socio-technical analysis brings about the issue how, and under which conditions technical objects and inanimate objects can become factors of the social world - more precisely: of social interaction. In order to find answers supporting the argument of continued development one needs a concept which is similar in approach to F. Braudel's (1979: 18) when he explored the "possibilities of a pre-industrialised world". For our purpose - to understand the possible development of nowadays industrialised/mechanised world - one has to base the sociological foundations from which emerge the technical machines in sound working order and which - then, once applied - seem "absolutely necessary".

"Mechanisation takes command" ?

S. Giedion (1948) provided lots of facts to foster this, his famous statement. In this article I will not argue against this proposition, yet I will ask for the societal background and for sociologically eminent reasons why mechanisation got and incessantly possesses this relevance. In which way, using terms and theoretical concepts can the mechanisation of the world be appropriately characterised and investigated? Besides, new models for interpreting technology, new theoretical components of sociological analysis and explanation are becoming required. Complementary reformulating requirements come to light in the area of sociological self-perception: Fruitful sociological investigation into technology is not possible without overcoming handed down sociological barriers. There is more to the sociological investigation into technical development than merely generalising the relationship between human subjects and material objects. An individualistic consideration of humans and machines allows no sociological conception. In this direct polarity, every indication of social ordering and forms of inter-human relationships is lacking. For social scientific research into technology, new entries have to be opened. This should help to understand the functioning of technical systems within society. This implies to refrain from the two-way separation of "technical" systems on the one hand and "social" systems on the other. Instead analysing the implementation, organisation and dynamics of "socio-technical systems" are to be placed in the limelight.

In socio-technical systems, technical and social elements are undeniably connected. Following this principle, in the socio-technical system of traffic for example, pure technical solutions strictly lead to undesired consequences whilst proposals for changing basic social conditions (disregarding technical elements) of the traffic system usually appear as "technically impractical" speculation.

If then "technology" cannot be investigated alone, what should the focus of sociological analysis of "socio-technical systems" be? It appears to be necessary to understand technology as something changeable within the equally transforming societal processes. Thus, a dynamic term for "technology" has to be placed at the centre of socio-technical analysis.

A dilemma comes to light similar to that stated by N. Elias (1987) regarding the formulation of the very question of sociology: It is not a matter of the opposition or sociologically determined difference between "individual" and "society". Both exist within each other. Every society is composed of individuals who, from their point of view, cannot exist as social beings without society. Hence Elias talks consequently of "the individual's society". Following along the same vein, in the sociology of technology the presumed and often enforced difference between technology and society is not compromised. To a certain extent, the social aspect is dependent on technology whilst from the opposite angle the technical aspect cannot be separated from its social components. As a consequence, it cannot be an issue of "technology and society" but rather an investigation into the technology of the society.

In the analysis of socio-technical systems, the term "mechanisation" comes closest to smoothing the way for the investigation of a special sociology of technology. It relates to social acting: It characterises the transference of an individual social action - in certain cases only from sequences of acting, in others from related series of actions - upon technical structures which are being ever extended in industrial societies of today. It is precisely this a culturally created objectivation of laid down social contexts in concretised or abstract structural components of these systems.

The structural components can thus become dominant. Observation of machines and appliances in a narrower sense provides data and grounds for empirical measurement of "grades of mechanisation". The current grades of mechanisation can be seen as indicators in which way and to what extent a society has made itself dependent on technical systems and processes.

The concept of the mechanisation of socio-cultural elements offers an explanation for certain social processes, although the question as to exactly how and why remains open. In a fragile social world, how does the situation arise that in social interactions complicated, numerously interlinked processes constantly recreate stable development trends, are able to consolidate and even gain long term dominance?

An approach is the concept of "formative principles" (J. Hochgerner, 1986). Having grown with history and standing strong in the industrially developed societies, the principles of "hierarchy", "objectification", and "growth" are at the heart of this theoretical framework. However these are not signs pretending to show the direction in which development will take, nor do they describe normative expectations referring to the historic dynamics of civilisation processes. Rather, they represent a practical implementation and, in large social systems, a specifically related further development of the microsociological term "figuration" (N. Elias, 1978: p. 142). Included in this are "transformation models" in which people act, according to the situation, "not only using their intellect, but also with their complete body, and every action in their relationship to each other".

Hierarchy, objectification, and growth are termed "formative" because they respectively bundle up a range of material and idealistic elements together with organising social facts (laws, customs and traditions, role expectations etc.) in such a way that the societal formation which is supported by this is in fact secured by their existence. This security of continued existence incorporates remodelling and possible change (throughout the society as well as even concerning its formative principles). For societies such as the industrially developed which recognise "growth" as a constitutive necessity, constant change even guarantees preservation: Such a societal formation can only have continued existence if it is in the position to remodel itself in a controlled way to incessant change. Such regulation, which controls human behaviour according to the specific needs of a certain given social formation, is termed a "formative principle". It organises existence and change in social behaviour over several historical eras without itself being connected to the respective form of that time.

Should technologies and elements of social acting merge into socio-technical systems, the technical components would be "moulded" according to the societally dominating principles. These moulding principles - in tracing the ideas of F. Braudel - provide "the societal foundations" from which the dynamics of mechanisation are determined and socio-technical systems gain structure and function. Whilst technology acts as an organising body within these systems and the organisation of social relationships changes itself due to the influence of the mechanisation, a "social organisation of technology" takes place.

The social shaping of technology and levels of mechanisation within a certain society cannot be derived by analysis of technological characteristics (performance, function, its composition and construction). It is not only individual innovations which cumulatively constitute technical progress, or to be more blunt determine social transformation. The moulding principles are responsible for "organising" existing and new technologies together with social, material and non-material norms or rules. Only in this form is technology effective as an agent of change in the most different applications in socio-technical systems.

Behind technical concepts for the solution of certain problems stand societally defined forms of awareness of concrete problems, as well as socio-culturally moulded presumptions about what the solution could in fact be. In this sense investigation has to follow the line of establishing how the moulding principles of hierarchy, objectification and growth manifest themselves in concrete technologies. Using new trends in the high-tech area, this can be observed. In this way, information and communication technologies as well as processes and concepts of biotechnology play a prominent role. The starting point is created by the question how technological transformation works in the development of specialised technical fields and whether or not societally moulding principles do in fact paradigmatically have their own part to play.

Transforming basic modes of technological progress - examples and patterns

Simply listing "old" and "new" specialised fields would be inadequate to explain the existence of a "technological transformation". Alongside traditional fields such as "mechanical engineering" and "civil engineering" exist completely new areas like "biotechnology", "nuclear physics" or "software engineering". The typologistically important changes only become evident when the internal dynamics within the "old" and "new" specialised fields are compared.

Such comparisons are possible at several levels. When making comparisons between traditional and new technical specialised fields, the most striking element is the subordination in the relevance of material substrata: Both technological work itself and its products were and are linked to circumstantial manifestations in mechanical engineering, mechanical technologies, civil engineering and similar technological specialised fields. In as far as the newer information and communication technologies, genetic engineering, informatics and artificial intelligence are concerned, manifestations of that kind play a lesser role. Immaterial performance and function frequently take the place of material products. Both the tactile experience of technical work as much as the perception of results tend to point from an originally rather concrete to an increasingly abstract "pole". On the greater scale of the practical economic application of these differently funded technologies corresponds to the trend of the transition from the traditional heavy industry to "meta-industry" (W. Wobbe, 1986).

A further differentiation between the specialised areas emerges when one considers the development of those technical processes, which come about in a typical way. A visual comparison of a gear wheel and the micro processor clearly accentuates this difference. Clearly both are applicable in different technical systems; whilst the function parameters of the gear wheel remains the same - because they are set out - the microchip can not only calculate in a more varied way with different inputs, it can also produce manifold outputs with corresponding diverse programmes. The more traditional technical specialised fields are characterised by the fact that the greater part of its processes run relatively rigidly and mechanically (sequential and consecutive). The newer technological applications however are to a much greater extent more complex and variable, and to some extent "autonomous" in their operation.

Within this transformation of the specialised technical areas, a shift from rather materially concrete, process oriented rigid technologies occurs towards rather immaterially, abstract technologies, having a much more complex and variable appearance.

However it is not possible to place the individual specialised areas exactly on a scale. This is a fundamental impossibility not merely a question of lacking data. Yet in the first approach, three groups of specialised technical areas can be differentiated as being more or less "traditional" or "advanced".

Belonging to the first group are all simple mechanical technologies, also assembled mechanical processing technologies, civil engineering and architecture, mechanical engineering, material science and diverse processes of exploiting materials. All of these branches indicate a relationship to material products and concrete production methods. For the most part they represent that classically encapsulated idea understood in the term "technical", and whose image is fostered by numerous "typical" technical professions. Architects and civil engineers, toolmakers, lock smiths, iron and steel workers and mechanical engineers in all specialist and professional areas constitute the central - formerly clear - job outlines of these dominant "mechanical specialised fields of technology".

Recently these images have changed. Whilst the self concepts of architects are being damaged by "mechanic" formalisation (which restricts creative freedom in material structures), technical advances are beginning to distinguish between abstract means of work and immaterial products. New construction, prospection and production technologies are merging from these at one time so graspable "real" fields of work into the system structures of the new "information and communication technologies". They are progressively assuming their operational characteristics and form.

In the nineteenth century when industrialisation was fully underway, electronic and chemical engineering, associated material sciences (from metal to artificial), as well as the natural scientific technical subjects assumed the deviant form of mechanic field. They laid the foundations for the so called "science based industries" and developed specific qualification demands for the engineers: Those being of relatively higher abstractness of working methods and processes encouraging school and university education. The role of science in being the impulse provider for technical and industrial development was exited from this specialised field. This second group is characterised in such a way that more emphasis is increasingly placed on research and development than on experience. Here, in this somewhat transitory area, the main issue is "specialised fields based on scientific knowledge and research".

The third group of specialised technical fields pushes the notion of basic scientific support and extension even further. Technology now gradually takes on a shape which relies very little on the existence and functioning of traditionally operating mechanical components. Paradigmatically stated, the inherited forms of "mechanic" technology can be compared with the most advanced fields, that is the so called "cerebral technologies". Included in this term are software technologies, as well as biotechnological processes as long as they exhibit elements of artificial intelligence (e. g. "open-access" machines and other automatic apparatus or processes with the ability to learn). Proximity, on the technical level, to the way in which the brain functions provides the precondition for achieving a kind of co-evolution of consciousness and machines (cf. J. C. Glenn, 1989, H. Moravec, 1993). Cerebral technologies are characterised by a dominantly immaterial character of their measurable components just as much as the operation produced. The "sphere of possibilities" - both in implementation as well in the intended and unintended consequences - greatly surpasses the corresponding extent of traditional machines, constructions and material processing. In this way the emerging technologies of this kind are facing a specialised, completely new challenge. This can be seen particularly in the trend towards determined abstractness of the course of events, which increasingly excludes judgement by - biologically and culturally - acquired sensoric perception. This indicates more than just the growing claims of cognitive, "brain centred" performance and ability, or in another way the vague hint of functioning in a way similar to that of the brain in certain technical machines and systems. In fact a fundamental transformation of technical basic elements and their functional scope of use is opening up which demands a new, conceptually comprehensive understanding.

Regarding everything which previously was understood as being technology, even in its broadest sense in everyday and scientific understanding, "mechanics" were the issue: Moving metal components, driving motors, flows of kinetic and thermal energy all constitute the traditional illusion of technology. In all already achieved precision being complex and diverse it is quite clear that the main components allow the description "mechanical" technology. But there is just as little doubt regarding F. J. Dyson's (1988) equally as extensive and different diagnosis of time. According to him, the emphasis will shift in the years to come from the current "metal-silicon-technology" to the "neuro-enzyme-technology" of tomorrow. This is best illustrated by a short representation of huge range of future technical developments in the "macro" and "micro" fields. Whilst on the one hand projects dealing with technological development are becoming bigger and greater, the much anticipated challenges of techno-scientific developments remain in the field of micro instruments and the technical triumphing of molecular to atomic and sub-atomic processes. Extremely large programmes and projects (space travel being a "classic" example and the construction of the computer industry) become relevant occupational fields for engineers.

Design, development and evaluation of especially extensive plans are becoming more and more important. These respective requirements are not comparable with the work of the traditional engineer in construction and projecting. Management functions and technical knowledge have to be connected in such a way within the framework of an appropriate cultural self appraisal that the division of work no longer only implies the distribution of available work, but leads to the creation of previously unknown results by means of the joining of all conceivable contributions ("macro engineering").

As enormous efforts made by the Japanese for the development of the "fifth computer generation" of intelligent systems shows, it is often a question of "micro technology" (in the sense of the size of products) in such macro processes. A still somehow traditional, mechanical image of technology shows evidence of a development path of the so-called "micro-tools". New and very specific engineering occupations of growing relevance (about the term and professions of "micro-engineering" cf. I. Amato, 1988) concern the finest mechanical parts and apparatus in the field of micrometer dimensions (a scale of one millionth of a metre). Application spheres are found mostly in medicine (intravenous to coronary blood vessels, robots to treat heart attacks; electrodes for epilepsy therapy) or in environment technology (highly sensitive gas chromatographs for air pollution analysis).

Whilst such developments - although still perceivable and therefore still inspirational - still remain in the future as far as the intended industry of use is concerned, the project conceptions of the one thousand times smaller "nanotechnology" extend much further. In this way technology reaches levels of molecular structures and transcends previous limits of technical processing of (any kind of) materials: Under the condition that atoms are no longer seen as being in "disordered groups", but are able to be directed and "organised", artificial molecules emerge which biochemically replicate themselves and are able to grow together with genetic technologically controlled functions in certain machines.

The biochemists of tomorrow construct motors and moving parts molecularly out of proteins. The fact that proteins are not able to fulfil all expectations of workshops implies that the indicated leap in the development of genetically engineered artificial materials will lead to the second generation of "nano-machines", so-called "assemblers". The first use of which will most likely be in the field of the continued miniaturisation and acceleration of computer technology ("molecular electronics" - the current does not flow along the most minute printed circuit but through molecular structures; specific, genetically engineered molecules will function as switch components). The perfecting of such assemblers in nanotechnology can revolutionise fundamental technology to such an extent that even Aristotle's differentiation between nature and artificially derived technical products may become invalid. According to Aristotle's previously undisputed example, the bamboo wood in a bed frame may - provided it is not organically dead - grow into a new piece of wood, but it never can grow another bed. Yet exactly this type of creating technology by means of technology should become possible according to the nanotechnology concepts of the assembler principle. To a certain extent that is the genetically pre-programmed self-creation of not only amorphously but also of functionally structured materials (cf. K. E. Drexler, 1986).

On the horizon of such developments lays the explanation potential of the collective term "cerebral technologies". The already outlined developmental lines concretely point toward technological development in this direction. Therefore it appears to be worth suggesting this expression as being the term for basic technical novelties, because in any case it envelopes the facts better and more comprehensively than do currently circulating terms such as "cybernetic" technology for example.

Alongside such fundamental trends in technical lines of development, new questions are being formed by the growth of synergetic impact chains. These are not merely longer, but above all are inaccessible to mechanic causal relationships of instructed, linear analytic thought. Meanwhile, the "classic" examples are the ever returning drastic examples which show the analysis of mistakes and behavioural processes in catastrophes resulting from uncontrollable large socio-technical systems. Accidents in nuclear power stations, chemical factories, traffic systems, or - extrapolation is justified here - in biotechnological procedures under the implementation of genetic engineering, all of which are not explained by the traditional image of the snapping of the weakest link in the chain. Increasingly, the functioning of the technical components in such systems is dependent on the perfecting of measurable shares of cerebral technology. Disturbances occurring in this way lead to a mistake or in other words automatically to a catastrophe, not because they are so unaware of human logic but often merely because they function asynchronically to the calculations of the service or maintenance personnel (cf. L. Hirschhorn 1984; C. Perrow 1984).

New technologies in the wake of formative principles

Observations of the current lines of technical development have been presented here in order to lead to an analysis of those societal preconditions which realise the great importance of technology for the organisation and transformation of social life in our times. Central in this context is the concept of the "formative principles" which is to be applied on the formation and effectiveness of technology. According to which, shapes and functions of new technology develop under certain conditions and criteria which are in a narrow relationship to the hierarchical foundations of societal organisation, to the objectification of social relationships and to the growth (in other words the acceleration) of productive and consumptive processes.

Within this general societal framework, technologies are being "organised". The constructive moulding of their functions, effects, and implementation follows a defined precondition. Both the integration as well as the consequences of technology upon the societal sphere affect this formation: The technologies created by social expectations and strategies coming from the protagonists (in this case individuals, groups, firms, public institutions, ...) become part of societal development in this "organised" form, and so appear as agents of change. "Transformation" can either mean adapting in the interests of securing existing foundations, or to extend away from a break in this development model. The second possibility is historically essentially more relevant and correspondingly more rare or in other words more unlikely. In this way, no successful adaptation of current trends of development to the dominant formative principles would arise, but they would themselves experience a reshaping.

The only question remaining is whether the outlined developments in the direction of cerebral technologies are to be evaluated rather as "principally" changing or as insistent trends (in the support of the dominant formative principles hierarchy, objectification and growth).

The increasing diffusion of components of cerebral technology strengthen the general function and efficiency of technology as a means of organising work, leisure, and private life as well. This is especially true for the information and communication technologies which organise information "objectively", arrange it hierarchically and intensify further information processing (growth). They intertwine individuals with social organisations in elementary forms of social interaction by means of the mechanisation of communication. Societal structures and civilisatory dynamics are thus modified in the direction of a general mechanisation. Since the character of technology changes, concomitant new socio-technical systems arise. According to the pace of spreading cerebral technologies, new ways of socially integrating these potentials will have to be found.

This process is based on the formative principles of the industrialised society being its organisatory mechanism. How far reaching change in the course of the current transformation processes is, and whether a transition to a new societal formation is underway, cannot reliably be stated here based on the considered information.

Quite clearly a change in the formative principles (in the sense of a new principle which could replace an existing one) is not expected of technology. Considered by themselves they only represent potential agents of change - with the greater likelihood of having an effect dynamically and primarily stabilising previous development trends. Only on the basis of their societal organisation - which is why their effects are essentially determined by the definition of this framework - can technologies become really effective agents of change.

In spite of successful conceptions and the implementation of (new) technologies under the supervision of social protagonists wanting to defend the status quo, technical resources will enhance existing relationships and further promote them. They will have a "changing" effect in the sense of an accelerated continuation of the previously drilled-in paths to increased mechanisation and greater social organisation dependence from the functioning of more and more complex and, at the same time, more and more narrowly combined socio-technical systems.

An alternative developmental path demands far more than the construction of new technologies. It necessarily implies guidance by innovation in the field of social organisation - thus particularly in that set of factors which regulate the existence and transformation of society (that is: the "formative principles"). Under such preconditions, trends other than those classically probable impacts (objectification, concentration of economic power through growth, new precisions in the hierarchically structuring in wider societal portions) would be feasible.

The fact that technology exists as a part of a self transforming society means that it is not a "pure technology" (in the sense of machines and working appliances) which society can bring to a new path of development. However, the complete opposite is also not to be expected that firstly new formative principles would be fully unfolded before another technology appears: Here, the issue is a common process of change in which the technical as well as the social components are organised (formatted) according to the same societal preconditions. Based on which, the analysis of this development as well as impulses to its management are able to fix on to either the social or the technical elements of existing socio-technical systems.

In order to fully understand and for the practical management of these processes a methodically clear delimitation of the field of investigation (a concrete socio-technical system) is required, as well as an exact as possible observation of the state of mechanisation (measurement or comparison of the achieved and of the sustained level of mechanisation), and an evaluation of the interdependent relationship between the socio-technical system and the formative principles. Procedures of studies in technology assessment (cf. J. F. Coates, 1974) can be drawn upon where limitations to this practical and politically oriented base would be produced with the briefly outlined theoretical and terminological instruments.

Regarding cerebral technologies, investigations of this kind should be carried out in an extension and concretisation of this outline. They would have to assume their approach by determining a field of use, that is: a socio-technical system in which they operate as technical components. In this way, answers to the questions raised in this outlined basis could be found. In particular long term effects of the societal as much as technological transformation currently underway are to be studied by applying such analysis to the progressive development and extension of cerebral technologies.

References

Braudel, Fernand, 1979: *Civilisation materielle, Economie et capitalisme*. Paris.

Coates, J. F., 1974: Some methods and techniques for comprehensive impact assessment; in: *Technological Forecasting and Social Change*, Vol. 6, pp. 341-357.

Drexler, K. Eric, 1986: Engines of Creation. *The Coming Era of Nanotechnology*. New York.

Dyson, Freeman J., 1988: *Infinite in All Directions*. New York.

Elias, Norbert, 1987: *Die Gesellschaft der Individuen*. Frankfurt/M.

Elias, Norbert, 1978: *Was ist Soziologie?* München.

Giedion, Sigfried, 1948: *Mechanisation Takes Command*. New York.

Glenn, Jerome C., 1989: Conscious Technology: The Co-Evolution of Mind and Machine; in: *The Futurist*, Vol. XXIII, No. 5/89, pp. 15-20.

Katsikides, S., Campbell, M., Hochgerner, J.(eds.), 1994: *Patterns of Social and Technological Change in Europe*. London.

Hirschhorn, Larry, 1984: Beyond Mechanisation. *Work and Technology in a Postindustrial Age*. Cambridge/Mass.

Hochgerner, Josef, 1986: Arbeit und Technik. *Einführung in die Techniksoziologie*. Stuttgart-Berlin.

Moravec, H., 1993: Geist ohne Körper - Visionen von der reinen Intelligenz; in: Kaiser, G./Matejovski, D./Fedrowitz, J. (eds.), *Kultur und Technik im 21. Jahrhundert*. Frankfurt/M.-New York.

Perrow, Charles, 1984: Normal Accidents. *Living with High-Risk* Technologies. New York.

Wobbe, Werner, 1986: Von der Fabrik-Industrie zur Meta-Industrie; in: W. Woebbe (ed.), *Menschen und Chips. Arbeitspolitik und Arbeitsgestaltung in der Fabrik der Zukunft*. Göttingen.

2 Critical Issues of Information Society

SAVVAS KATSIKIDES
DEPARTMENT OF SOCIAL AND POLITICAL SCIENCE
UNIVERSITY OF CYPRUS

Introduction

Though this paper will focus on different issues of information society, it is essential to recall some thoughts from other works in order to outline more precisely theory and empirical evidence.

In a recent published work (Katsikides, 1997), which has attempted to show that there is a variety of theoretical issues which concern mainstream sociology of technology a common understanding which derives from research in this field indicates that most sociological studies on technology use the comparative method, whereas the remaining surveys rather apply to the field of technology assessment. We have already argued that technology reflects the synergy of power and societal processes, and the latter must be analysed under the foci of sociology of science or even of the emerging sociology of information. While sociology of information should address a variety of theoretical perspectives that can be directed towards the social phenomenon of information, they alone do not provide sufficient insights into the nature of information either as an object of disciplinary discourse, or, as an object of nature (Balnaves 1993:108).

The emerging approach is that an entirely new concept is required and that there is a vital need for improved analysis with respect to the assessment of technological issues. It can be argued that theoretical considerations have to be linked with practical methodology in order to evaluate technological and societal issues, as different sets of complexities exist between the cultural and the operational aspects of the functional role of technology. However the issue here is more complex, and the argument can be summarised as follows. The first problem relates to methodology, where it is clear that a global approach, whether theoretical or empirical, reaches its limits very quickly. The second problem is a more general issue that refers to all the social sciences: a common direction to resolve common social phenomena is lacking. Thirdly, it can be argued that a new approach is needed, which would focus on a detailed evaluation and provide a synthesis of all the intervening variables involved in the technological discussion. One example of such an approach is the ARS model (Katsikides 1994). Finally, technological developments, like other social, economic, and technical approaches, are not socially neutral, and in the end they deal with different traditions (European, US, Scandinavian, Japanese, etc.). As such they accumulate social processes and reflect them, or, as Thomas Kuhn(1970) put it "a failure to assimilate fully new conditions and technology will strain the existing structures" of society. Existing structures, means the certain sociopolitical and economic span will continue in a sense to be constructing the system, where modernisation has still a long way to go. If we can see technology as a social phenomenon, by terms of sociological analysis and as such determined, then is socially constructed. This however, is definitely a case of socialisation and thereby could be socially "taught and learned." Questions, however, that could make people rethink their choices concerning technological change and problem solving, were in the spectrum of another work which was published in the Journal Innovation, in 1997, by the author.

Problem solving is a new parameter in the technological discussion and creates possibilities for redesigning the workplace. Further it offer solutions, which arise through the adoption of information systems and their impact on labour. Another key aspect in this work are the case studies in French hospitals. (Katsikides/Schneider 1994). In this work the problem arose in different hospitals and various models were proposed and finally adopted. The hospital sites analysed in the survey were selected from the public and private sectors. For the most part, they were attached to a university teaching hospital centre. All institutions had made or were planning to make investments in computing equipment, that besides handling financial management tasks, could also provide support for the organisation of a patient's hospitalisation from the perspective of a medical-technician or clinician. The study focussed on the information system development and acquisition policies applied, affecting the main characteristics of the target system as well as user involvement. Four hospital institutions are discussed. The first case is a paediatric city clinic that runs with an integrated hospital information system that was developed externally by a major computer manufacturer. The second case is a private institution for the treatment of cancer diseases, equipped with an internally developed system. The third case is the surgical department of a cardiological hospital operating with a highly complex local system combining the time-critical demands of the ICU setting with the general care requirements of pre- and postoperative treatment in the ward. The fourth case is a regional hospital centre maintaining a loosely coupled decentralised architecture of local networks.

Hulin and Roznowski (1985:47) in their work 'Organisational Technologies: Effects on Organisations Characteristics and Individuals Responses' define technology as the "physical" combined with the "intellectual" of knowledge processes by which materials in some form are transformed into outputs used by another organisation or subsystem within the same organisation. Law (Law 1987:115) for instance, saw technology as a family of methods for associating and channelling other entities and forces, both human and nonhuman. It is a method, for the conduct of heterogeneous engineering, for the construction of a relatively stable system of related bits and pieces with emergent properties in a hostile or indifferent environment. Berniker(Berniker 1987:10) claimed that technology refers to a body of knowledge about the means by which *we work on the world, our arts and our methods.* Essentially, it is knowledge about the cause and effect relations of our actions. Technology is knowledge that can be studied, codified, and taught to others.

Weick (1990:2) for instance, defines technology as equivogue that is something that admits of several possible or plausible interpretations and therefore can be esoteric, subject to misunderstandings, uncertain, complex and recondite. He provides a context that illustrates the strengths and weaknesses of prevailing thought about technology. Weick provides three definitions of technology. After reviewing those, we begin our analysis of different social scientific approaches towards technology.

The first definition, however, in Weicks understanding, underlines the explicit mention which is made of raw materials and a transformation process, items that are often implicit in other definitions. The new in this definition is the mention, that output might be used within the same organisation. It is possible, Weick, stated, that diverse technologies within the same organisation could exist. Finally, this definition is noteworthy because of its emphasis on processes rather that on static knowledge, skills and equipment. By equating technology with process, Hulin and Roznowski alert us to the importance of changes over time and sequence. (Weick, ibid).

The German Philosopher Heinz Hülsmann, (1916-1993) in his well discussed book "Die technologische Formation" (1985), which has not yet been translated into English, pointed out, when dealing with the idea of technology and its link to formation:

Technik ist nicht nur Können und Kunst, Fertigkeit und Wissen, sondern eben auch List und damit Gewalt und Herrschaft, wie dieses im Verhältnis der Form zur Materie sichtbar wird und wirksam ist.

My translation:

Technology is not only ability and culture, proficiency and knowledge, but also slyness and therefore violence and domination as their relation can be seen and gain efficiency to form and subject.

Hülsmann stated clearly the direction which sociological work takes when examining technology and its shaping. Although his book was published in the mid 80's it underlines the distinction between the European and the American context, which we will try to distinguish in our discussion. Salzman (1994) noted in his work, that the Scandinavian approaches, for instance could provide a model for ways to reorient design but also showed that their particular approach reflects a particular industrial culture (socialisation of workers, engineers, etc.) shaped by, among other factors, a longstanding craft tradition, a workplace environment of 90 percent unionisation, and requirements for labour union participation in many basic decisions that would be considered management prerogatives in the United States. A further point that has been made by D. Jahn (Jahn 1994), indicates the differences between the German and the Swedish trade unions; his findings demonstrate the differences in both countries. In Sweden, in contrast to the assumption expressed in the 1970s, technological progress has not been challenged to a larger degree. As a consequence there was no increasing tension between labour and capital or within the trade union movement.

A further definition of technology which was made by Law (Law 1987) and analysed by Weick, implies that the design and operation of technology do share some of those qualities but they do not exhaust the character of the processes it is unfolded in politicised organisations, and the above mentioned definition allows us to describe technology in a way more compatible with this quality of organisations, (Weick 1990:4). The third and last definition by Berniker, states that every technical system embodies a technology. It derives from a large body of knowledge which provides the basis for design decisions. Weick (ibid) in his own analysis argued that the first definition forces us to re-examine our knowledge of cause/effect relations in human actions and the choice of a different combination of machines, equipment and methods to produce the outcomes for which new technologies are instrumental. He continues, pointing out Berniker's argument and stating that technology follows rather than precedes a technical system, and, furthermore, that technology is both an a *posteriori* product of lessons learned while implementing a specific technical system and an a *priori* source of options that can be realised in a specific technical system. All of these analyses respond only partly, to the different debates and approaches when mapping technological and sociological knowledge. Furthermore, as Weick illustrates, other works by Scott (1987), Hancock,

Macy and Peterson (1983) and Perrow (1986), provide helpful summaries when definitions of technology have been translated into survey items intended to capture variations in skills, equipment, and technique. On this issue, Weick concludes that new technologies introduce a set of issues that organisational theorists have yet to grapple with. Unless they do so, the power of technology as a predictor of organisational functioning will diminish.

As new organisations are concerned with the introduction and useful usage of new technology, it seems that old industries or other enterprises can neither be compatible as previously Law (Law 1987) stated, nor follow a certain new form of organisational change, which is influenced by external parameters. Emerging technologies force both the organisational structure and external relations to become more efficient and proceed to optimisation. R. Reddy(1990:249) analysing the new forms of organisation concerning the US car manufacturing industry, which have to be more efficient, concludes that:

> ...if the United States continues to take four to five years to introduce a new car while Japan can do it in half the time, then obviously Japan will continue to increase its market share.

Let us recall the three logo model (LM), which has been discussed in Katsikides (Katsikides, 1997), and consider their application. A further point which can be made is the diversification between the European and the US technological concepts. The European concept is more state dominated (EU control) and several options and regulations, which have been approved by the Commission, are for instance, now not allowing subventions from the fifteen states and their regional local governments to industrial and other business conglomerations. The policies of these conglomerates against the EU regulations can be viewed each year on the list of firms which have to pay a penalty for ignoring these rules, or even for violating them.

The first perspective of the LM concerns the organisational structure, which can be found in every enterprise or organisation. The second, extrovertial, perspective shows the sphere of action of the organisation on the outside world. The third, introversial perspective covers the sphere of action in the inner life of the organisation, where all processes within the organisation are functioning. As we have seen, the structural change in organisations covers more and more enterprises. Public and private organisations are building their organisational structure on a common criterion. As a functional connection of all management and administrative starting points in the inner life of organisations apart from their size, joint criteria can be observed. The observation of this coherence will be shown in a shaping LM where the minimum of an organisational structure will be taken as a basis for new technology as the next logical path of other forms. The results of the various adaptations and applications of information technology, i.e. in production such as CAD, CIM, CAM etc. lead to the argument that a) automation and rationalisation effects were the first issue and b) the final result was the flexible oriented production. All these systems require a new method of administration. A new theme of organisation that includes fields and approaches such as data transmission, telecommunication, innovation of production, rationalisation of working operation, of employers etc. is now emerging. R. Kling (Kling 1994) states on this particular point: many organisations are adopting computing equipment much more rapidly than they understand how to organise positive forms of social life around it. However, some fervent advocates of computerisation portray the actual pace of computerisation in schools, offices, factories, and homes as slower than they wish. These "computer revolutionaries" argue that many key institutions - such as schools, businesses, family life, public agencies - can be progressively reformed through the appropriate application of computer-based systems. It should be now clear that changes in organisations might deliver the reason for other compatible tasks, which are included in the planning and might be adopted and implemented later. Hartmann (1986:180) makes the point that the crisis of the administrative work forms the compulsion of continuing production. The second point concerns the administrative operations that must go faster when necessary. The third point is the assistance to the administrative operation taking in account the flexibility of the enterprise; that means a faster collection and distribution of concrete transformations of information and data. It is not surprising therefore, that the installed system which creates new organisational structures, was established as an instrument operating regardless of social and political decisions. This implies that the organisation is not in a position to control the system further. The decisional parameters lie outside their action fields.

The second concept after the structural perspectives of the LM is to be seen on the external action radius of the organisation. The organisation analyses the relationships of the enterprises with the world outside. Pattern examples are mother and daughter companies, the relations with the state, the law, trade unions and other interested organisations and last but not least the customer. The compatibility of an enterprise is obvious through the synergies in the level of employment, on the concepts management and, finally on the sales and marketing area.

The third observation concerns the action sphere which is to be found in the internal structure of the enterprise or organisation. At this point we analyse trade unions, content of work, working time, collective bargaining, agreements, security, creation of work, economics, etc. The question that now arises is whether the US and the European technological concepts could in same way be compared and measured, according to the guidelines of European Union for the creation of a United Europe?

Further analysis has been undertaken by Egger E. (1995), in her book "CSCW: The bargaining aspect." From the informatics designing point of view, Egger has carried out a set of surveys which contribute to Computer Supported Co-operative Work (CSCW), that is, it contributes to the development of technical systems which at the same time are supportive of the work of groups. The innovative idea here is that CSCW integrates tools for working teams. This implies the creation of new measurements for work based also on the social environment. Through CSCW, which could become a vital issue for sociologists as Egger (1996:13) pointed out, in examining not only obvious technological systemic aspects but also the role of sociology. CSCW claims to support the working processes of groups. Therefore it is an interesting issue to analyse how people work together and which aspects of group dynamics have to be considered when designing technical systems, especially those focusing on planning processes. Therefore the contribution of sociology in defining design guidelines for CSCW systems has to cover working groups and group processes, bargaining situations and the role of time in groups. Controlling and monitoring were the first goals of the first generation of computer technologies, which have been implemented in production processes. Sydow (1985) used the term "joint optimisation" to describe the idea of combining the advantages of technical systems and human qualifications. Later the new idea which was to match principles stemming from the movement of industrial democracy, industrial psychology and usability of technical systems, was carried out by Kling (1984), Olson (1983) and Shaiken (1985). Bannon and Schmidt (1991), trying to define the work process, stated that, "co-operative work is constituted by the work processes that are related as to content, that is, processes pertaining to the production of a particular product or service." Holand and Danielson (1989) suggest three kinds of perspectives of Cupertino (see Egger 1996:27).

a) Cupertino as a strategy: Cupertino is based on solving disagreements and conflicts where the participants state their position very clearly. In the necessary processes of solving conflicts some participants try to persuade the others or try to push their interests by building coalitions.

b) Cupertino as co-ordination: Cupertino is described as a method for a group to solve some joint problem, or perform a common task. The process of Cupertino is based on sharing- among all participants involved- of the responsibility of reaching the goal.

c) Cupertino as reflection and creativity: Cupertino is interpreted as a group process where the partners are encouraged to contemplate and reflect on the matters being discussed. All participants are described as being potentially of equal interest for the observer.

Bannon and Schmidt (1991) in their work CSCW: Four Characters in Search of a Context, define CSCW as work by multiple active subjects sharing a common object, supported by information technology. A common object of work is to draw a need to a "shared goal" which is criticised as being too restrictive and to "shared material" which is criticised as being too loose. (see Egger:28).

Conclusion

D. Edge (Edge 1995:15) has pointed out two interesting approaches concerning the social shaping of technology. The first deals with the history and sociology of science and goes back to Pinch and Bijker(1984). The content of this conception is to study the development of technological fields, whereby it is essential to identify points of contingency or interpretative flexibility, where at the time ambiguities are present. When such "branch" points have been identified, the researcher then seeks to explain why one interpretation rather than another succeeded or why one way of designing an artefact triumphed over others. The second approach works "in" from the context. Here the starting point is the particular social context within which technical change takes place. The focus is on everything which contribute to shape technology. P. Senker (Senker 1995), makes an interesting point concerning IT and the role of the developing countries in perspective.

H. Mackay (1995:41) in his contribution "Theorising the IT/Society Relationship" gives an overview of the sociology of technology, pointing out that sociologists have until recently tended to avoid technology, but that this began to change significantly in the late 1980s with the growth and development of both (physical) IT and the (social) debate surrounding it. Mackay is on the right path, when he stated that:

> sociologists of technology are concerned with explaining how social processes, actions and structures relate to technology; and in this are concerned with developing critiques of notions of technological determinism. The theories and concepts which have been developed

are increasingly recognised as of value to technologists, notably in the area of information systems design.

If technological determinism implies that the concept that technology is autonomous, then symptomatic technology as stated by R. Williams (Williams 1974) explains the contrary, namely, that technology is a symptom of social change. Mackey further, (Mackey 1995:41), explaining this point, stated that "according to this model, it is quite clearly society which is in the driving seat of history: given a strong social demand then a suitable technology will be found".

Although from lab invention to wide market consumption it might sometimes take up to seventeen years, (Profil 1997:19) it should be noted here, that other socio-political and economic factors play a vital role in determining whether a certain technology is adopted or not. Braveman (1984) arguing about technology, also makes the point that technology can not be focussed only on individual inventions. Furthermore, it is crucial to examine how broader socio-economic forces affect the nature of technological problems and solutions (see also Mackey 1995: 43).

References and Notes

Balnaves, M. (1993), The Sociology of Information, *ANZJS* Vol. 29, No.1, March 1993, p.108.

Bannon L. and Schmidt K. (1991), CSCW: Four Characters in Search of a Context. In: Bowers J.M. and Benford S.D. (eds.), *Studies in Computer Supported Co-operative Work,* North Holland, Elsevier, Amsterdam.

Berniker, E. (1987), Understanding Technical Systems, Paper presented at the Symposium on Management Training Programms: Implications of New Technologies, Geneva, Switzerland, Nov.1987, p.10, also cited by Karl E.Weick, Technology as Equivoque: Sensemaking in New Technologies. In Goodman S.P. et al. (eds.), *Technology and Organisations*, Jossey-Bass Publishers, San Francisco, Oxford, 1990, p.3.

Edge, D. (1995), The Social Shaping of Technology. In: N.Heap, et al. *Information Technology and Society*, A Reader. Sage.

For Britain, Edge notes (1995:28). "How, then, might Britain's performance be improved in this respect? How can we increase attention to the entry into use, and more widespread adoption, of new technologies, and to strengthen the influence of the exigencies of these processes on the generation of new "basic" knowledge, and the development and design of new products? exhortations to "so better", or to be "more like the Japanese", are simply not going to be enough. For, typically, the different types of knowledge involved in the different phases of the product-cycle model are possessed by different types of people, and there is hierarchy of prestige that rises as we move from right to left across our diagram. "Basic research" is more prestigious than the mundane -but commercially crucial- tasks of implementation, marketing, distribution, maintenance and repair. All these involve technological knowledge, but it is knowledge our culture typically devalues".

Egger, E. (1996), *CSCW: The Bargaining Aspect*. Vienna/Frankfurt/New York: Peter Lang.

Giddens, A. (1976), *New Rules of Sociological Method*, London.

Giddens, A. (1982), *Profiles and Critiques in Social Theory*; Berkeley, University of California Press.

Giddens, A. (1984), The Constitution of Society. *Outline of the Theory of Structuration*, Cambridge.

Hancock, W.M./Macy, B.A./Peterson, S. (1983), Assessment of Technologies and Their Utilisation. In S.E. Seachore, E.E. Lawler III, P.H. Mirvis and C. Cammann (eds.), *Assessing Organisational Change*. New York: Wiley.

Hartmann, M.(1986), Strategien und Resultate der Verwaltungs-rationaliserung. In: *Journal für Sozialforschung*, Heft 2, p.180.

Holand U./Danielson T. (1989), The Psychology of Cupertino- Consequences Descriptions, *TF-Report* 58/89, Oslo.

Hulin, C.L. and Roznowski, M.(1985), "Organisational Technologies: Effects on Organisations: Characteristics and Individuals Responses." In: L.L. Cummins/B.M. Staw, (eds.), Research in *Organisational Behaviour*. Vol.7. Greenwich, Conn. JAI Press, p.47.

Jahn, D. (1994), The Challenge of Technological Progress in Modern Societies. In: Katsikides, S. et al. (1994), *Patterns of Social and Technological Change in Europe*. Avebury, Aldershot. Jahn's point: When looking at Germany, we see that in the dimension of productionist/alternative world views the gap between capital and labour has widened. Since the compromise on societal development oriented towards technological progress and economic growth is a basic principle for industrial society, this gap may lead to the conclusion that there are some hints for an undermining of the societal consensus. However, this cleavage does not run very nicely between labour and capital, but instead it shows that there is a cleavage between parts of labour, on the one hand, and capital and other parts of labour, on the other. This conclusion even leads to the fact that we may expect a rising cleavage within the labour movement. This cleavage runs along the lines of "radical" versus "moderate" factions in the labour movement. The early 1990s have become challenging times for the ideas that criticise productionism as economic security and nationality has reasserted itself throughout Europe. The revolutionary processes in Eastern Europe and the unification of the Federal Republic of Germany may have led to a decrease in the importance of the cleavage analysed in this article. Particularly the standpoints of the proponents of technological progress have been supported by this development and material aspects gained in importance. At the moment, it is difficult to predict whether or not this trend will continue or if the opposition against the dominant industrial culture will become alive again. However, the form and content of such protests can change significantly in that there is a movement from a protest that emphasises the irrationality of social development and fragmentation of culture by stressing egalitarian values, towards a protest that hopes to interrupt the process of alienation by marginalisation of underprivileged groups and stressing national values".

Olson, M. (1983), Remote Office Work: Changing Work Patterns in Space and Time. *Communication of the ACM*, 26(3) March, pp. 182.

Katsikides, S. (1997), *The Societal Impact of Technology*, Ashgate, UK.

Katsikides, S. (1994), Interests in the Transformation of Organisations. In: Katsikides, S. (ed.), *Informatics, Organisation and Society*, Oldenbourg Verlag, Wien- München, p.47.

Kling, R. (1984), Assimilating Social Values in *Computer-based Technologies, Telecommunications Policy* (June), pp.127.

Kling, R. (1984), Usability versus Computability: Social Analyses by Computer Scientists. In: Katsikides, S. (ed.), *Informatics, Organisation and Society*, Oldenbourg Verlag, Wien- München.

Kling states further (1994...) as Perspective four: Professional Responsibility to Society is Essential, in the above mentioned paper, the following: Some computer specialists are specially concerned that computing technologies should be "sound products." According to this view the computing professions should be responsible to their clients by delivering practical systems which are usable, reliable, and safe. The main foci of attention have been to identify computer systems which can be major threats to physical safety or civil life and to identify discrete solutions to reduce these risks. Some forms of computer technologies can be harmful because of unreliable software (e.g., life-critical information systems; election counting systems; social security payments; fly-by-wire aircraft; military command and control systems.). Other kinds of computer based systems threaten to diminish personal privacy. Both kinds of threats have been the subject of a specialised topical literature, the concern of organisations like Computer Professionals for Social Responsibility and special forums, like the ACM sponsored "Risks" computer bulletin board. In this perspective the primary reforms will come through improved software quality and certain changes in organisational practices (e.g., privacy protections; administrative guidelines to insure safe software and data handling practices).

Kuhn, Th. (1982), *The Structure of Scientific Revolutions*, Chicago University Press.

Law, J. (1987), Technology and Heterogeneous Engineering: The Case of Portuguese Expansion. In W.E.Bijker, T.P.Hughes, and T.J.Pinch (eds.*)*, *The Social Construction of Technological Systems*. Cambridge, Mass.: MIT Press, p.115.

McLellan, ed. (1988) *Marxism: Essential Writings*; Oxford University Press.

Murphy, L. (1994), Technology and Society: An Introduction to the Political Sociology of Organisation". In: Katsikides, S. (ed.), *Informatics, Organisation and Society*, Oldenbourg Verlag, Wien- München.

In his interesting contribution Murphy notes the following about H. Marcuse:"The famous analysis of the use of technology in manipulating social consciousness is Hebert Marcuse's One Dimensional Man (1964): "The technological and political conquests of the transcending factors in human existence, so characteristic of advanced industrial civilisation, ...exerts itself in the intellectual sphere: satisfaction in a way which generates submission and weakens the rationality of protest". (in, McLellan 1988:353). Marcuse sought to show that in the immediate post-war era, when science and technology were heralded as having found solutions to social and economic problems and subsequently negated political conflict ('end of ideology'), the "absorption of clashing opinions, of the driving power of negation, into technological rationality actually means that 'advanced industrial culture is more ideological than its predecessor'".Anthony Giddens (see also Giddens 1982:152-153).

For Marcuse the organisation of capitalist mechanisms of production no longer supply the explanatory power needed to analyse major institutions. In contemporary society the consciousness of the individual is subject to state tolerated suppression; in other words, that social freedom and organisation for the individual are tolerated so long as the individual does not confront economic organisation. This view of the state, organisation, technology, society, and the individual is expressed by a member of the "Frankfurt School of Social Research" 'in exile', a group of Marxist social theorists who witnessed the rise and rule of the Nazi regime in Germany. The experience of fascist ideology in Europe is an illustration, par excellence, of the way in which the state can use organisation and technology to manipulate the consciousness of a mass society.

Perrow, C. (1986), *Complex Organisations*. 3rd ed. New York: Random House.

Profil. (1997), *Technische Revolution*, Nr. 27, 30 Juni1997, 28 Jg. e19, Wien. Austria.

Reddy, R. (1990), A Technological Perspective on New Forms of Organisations. In: P.S. Goodman et al. (1990), *Technology and Organisations*, Jossey-Bass, San Francisco, Oxford, p. 249.

Scott W.R. (1987), *Organisations: Rational, National and Open Systems*. Englewood Cliffs, N.J.: Prentice-Hall.

Senker, P. (1995), Technological Change and the Future of Work. In: Heap N et al. (1995*), Information Technology and Society, A Reader*, Sage. p. 146.

Senker makes an interesting note on the role of IT, its perspectives and the role of the developing countries. He notes: Japan shows every sign of maintaining its technological lead in IT and advanced materials, with the USA following, and, perhaps, a growing gap between Japan and the USA on the one hand and Europe on the other. New Technologies are likely to be exploited mainly to provide better paid and more interesting jobs in the countries which dominate their initial development and exploitation. The main benefits in terms of production and development are likely to accrue to the more advanced countries-in the next few decades a category which may include the advanced regions of some newcomers such as South Korea and Brazil. Nevertheless, many developing countries seem likely to fare relatively badly. Biotechnology, for instance, is likely to increase rapidly in significance in the early years of the twenty -first century. It is unlikely that developing countries will benefit disproportionately in terms of employment: indeed, such trends as it is possible to discern indicate that advanced countries may well benefit at the expense of developing countries.

Shaiken, H. (1985), The Automated Factory: Vision and Reality, *Technology Review* 1/85, Washington.

Williams R.(1974), Television Technology and Cultural Form, London, Fontana.14. (see Noel Perrin. Giving Up the Gun. Japanãs Reversion to the Sword, 1453-1879, David R. Godine, Publishers, Inc. Boston, Massachusetts, 1989).

3 Culture Gap or Culture Trap?

JANE YOUNG
LEEDS BUSINESS SCHOOL,
LEEDS, ENGLAND

Introduction

The computer power which is at the heart of modern IT is just reaching its half century. An appropriate time perhaps to reflect on the issues which surround this technology and the Information Systems (IS) it enables, which have such a major impact on the workings of all organisations, and the lives of those who toil within them.

As IS & IT have become more central to the success of organisations (integrating IT into the business strategy is seen as the second most important issue for IT managers in the 91/92 Price Waterhouse IT Review) so the need to understand the factors which impact on the success of IS throughout their lifecycle increases.

In the early days of the computer age, organisational use of Information Systems was in promoting efficiency, automating existing tasks. As time passed the advent of the management information system added a new dimension, providing systems which assisted managers by making available much more information to enhance their decision making. Used wisely these systems were a useful asset, but the wisdom of Ackoff in his description of 'Management Misinformation Systems' (Ackoff, 1967) should not be forgotten, an initial glimmer of situations in which systems are perhaps not all they seem.

The eighties saw the arrival of the strategic information system - systems which supposedly provide the competitive advantage so eagerly sought by organisations influenced by Michael Porter (Porter & Millar, 1985). There are many examples of these systems which gave the edge over the competition (see Madnick 1987, Ward et al. 1990, or Galliers 1991). Thinking about the role that IS could play, developed further until the concept of a Strategic Information System (SIS), was extended to include those systems which directly supported an organisation's business strategy (Galliers 1991). This was a good time for IS people, they were now being seen as central players in developing and enabling the organisations' corporate strategy, they are starting to move out of the technological ivory tower and into the boardroom.

The status of IT professionals improved further when 'Reengineering the Corporation' from Hammer & Champy became a best seller, they were seen once more as being at the forefront of organisational innovation. Whatever the merit of the book and the real benefits obtained from the systems described within its case studies, the idea or concept of re-engineering with its perception of rebirth, fired the imagination of managers seeking a new 'comfort blanket' as they struggle for support to survive in an increasingly turbulent and hostile, business environment. Making radical improvements in business performance using IT, became top of the management agenda.

The systems advocated under this 'don't automate, obliterate' (Hammer 1990) strategy were supposedly capable of supporting radical change in the organisational processes, with associated benefits. There must surely also be radical change in the organisation itself, some of it planned and intended, some unplanned and emergent. Newton's law of:- 'every action having an equal and opposite reaction', in an organisational context.

In terms of organisational change, Venkatraman (1991) envisages further possibilities still, suggesting that the ultimate goal is that of 'business scope redefinition'. The organisation will be transformed and reconfigured using IT.

The theory therefore suggests that glittering prizes are available, yet practice is very slow in delivering; Galliers(1991) admits that many strategic information systems are 'happy accidents' rather than part of a planned strategy. It is claimed that 70% of reengineering initiatives are failures (quoted at the 1994, Cranfield BPR conference). Venkatraman sees reconfiguration as a long term goal rather than an immediate reality.

Using information systems

This is the fundamental dichotomy, to consider, some organisational information systems, whatever their role are extremely successful, with clear benefits (see Ormerod 1994 describing the situation at Sainsbury's), while others are equally unsuccessful in that they not only do not perform as expected, but have damaging consequences (London Ambulance case quoted in Which Computer, December 1992, Taurus quoted in Computer Weekly, 1992).

The tension increases as the demands from within the organisation for the possible benefits from information systems increases, but IS professionals are painfully aware that IS enactment is not a deterministic process. Understanding what contributes to success and what contributed to a perceived failure is a critical area of enquiry. In addition the role of IS within organisations is not clearly defined or understood, this must also be addressed.

Perhaps the role that is played by IS in organisations is the role that they are allowed to play. Is it realistic to assume that any particular system plays a defined role? Is a system's role actually defined by the circumstances in which it is conceived, decided upon, developed, implemented and assessed, i.e. the whole enactment cycle? At the point of initiation there will be a unique set of factors or parameters operating within the organisation, within its market sector, within the global business and technological environment - almost the birth chart for the new-born system. These parameters will shape its growth and development - stony ground from the unenthusiastic or intransigent, and growth is stunted, warmth and enthusiasm gives food and nourishment and a healthy child develops. The way in which the immature system is introduced to its foster family is critical. Does it look like them? not such a silly question - looking and feeling 'comfortable' to the foster family is crucial. Will it fit in? Will they have to change - is it intended to make them change? The system needs a 'friend' to speak up for it - to fight its battles to see that it settles down and starts to fill a useful role. Done well and a healthy infant grows, performs well and fulfils expectations. Not done well and the reverse occurs - a misfit system that struggles and fails to fulfil expectations, it also creates a poor perception in its' space, of the capability of such systems, a perception that can become sedimented into the organisational mindset, making it incredibly hard to change, and creating a difficult organisational climate for IS in general.

The essence of understanding this situation, lies in understanding the dynamic interaction which is taking place, and systems thinking is ideally suited to this task. As Jay Forrester suggests, 'Understanding and managing change are central tasks in both technological and social systems' (Forrester, 1994).

Role of culture

The need for an IS to 'fit' has been mentioned - what does this mean in organisational terms. The term generally used to define that 'je ne sais quoi' which makes organisations special is culture. The working definition of culture used by Checkland (1990) is that it is the 'roles, values and norms' which prevail and make each organisation different. Senge (1990) puts it more simply still and says that culture is 'the way we do things round here', implying that the determinants of culture are part of tacit organisational knowledge. So for a system to fit it must be congruent with the organisational culture, and the culture must be sufficiently flexible to cope with the innovation that the system enactment will bring. Understanding what this means is of great importance for all organisations. It has been said that one major reason for the acquisition of inappropriate, or ineffective information systems is the 'culture gap' between the IS professionals and the organisational managers.

What is the culture gap ?

The culture gap is the difference in thinking regarding IT, which is dependant on the background of those involved. Grindley (1992) in the Price Waterhouse IT Review of 1991/1992 describes the situation. On one side is the business manager who does not appreciate the potential contribution that IT and IS could make to the business, resulting in lost opportunities. On the other side is the IT specialist who sees great opportunities, but fails to understand the impact that the proposed technology and systems might have on the people who make up the business.

Grindley also suggests that often the IT specialists are felt to belong to some invisible 'computer university', feeling more fellowship with other IT specialists than with their business colleagues. The potential for this syndrome to increase, is more likely than ever, now that the information super-highways of the Internet, and World Wide Web are with us. For many technical specialists 'surfing the net' is so much easier, than trying to communicate face to face, with real people.

There are those who think the problem is exaggerated, five per cent of the Price Waterhouse sample felt that the culture gap, in relation to IT specialists, is no more evident that with other specialist functions, such as finance, engineering or marketing. That does however, leave ninety five per cent who do see it as a problem, forty seven per cent of the IT directors surveyed felt it was their major problem.

This is of necessity, a simplistic picture of the situation, and one that has changed since Grindley's description in 1991. The huge amount of hype surrounding business process reengineering has meant that business managers are now very aware that IT has been used very successfully by some businesses. Many are keen to jump onto the bandwagon, encouraged by consultants and IT specialists alike, but have little idea of what they might be being persuaded to espouse. Similarly the perspective of the IT specialist may not be unbiased or holistic, or so it seems to those of us with a cynical world-view, as the claims made for the curative properties of this magic IT cure for ailing businesses, bears a strong resemblance to the claims made for the philosophers stone. (Jackson (1993) echoes similar sentiments).

The fact remains that there is still a large gap between the perceptions of the business manager and those of the IT specialist, with regard to the role and place of IT & IS in business.

An important part of this is that the IT specialist, who is completely at home with the technology, fails to appreciate the fears of those for whom this new organisational tool is merely a powerful new threat to 'the way we do things round here'. A threat because it is outside of their experience and knowledge so that they are unable to comprehend the new shape their world might take when this technology arrives. It is no more than fear of the unknown, but this has been a powerful influence since time immemorial. This potent force acts to maintain the status quo, even when this is itself problematic, how else can one explain the reluctance of an electorate to vote out an oppressive regime when the opportunity presents, hoping instead that things will improve.

Bridging the gap and the unintended consequences

Now to consider a case study of an attempt to bridge this gap in perceptions, and improve communication, which suffered from the unintended consequences that so often follow attempts at action in human activity systems.(see also Forrester 1971 & Senge 1990).

Case Study - The organisation in which the information system exists consists of a central management sector, known as 'the centre', and five divisions, each of which has its own management team. The management information system being studied, which will be referred to as (CMIS), is part of a suite of systems. The trigger for development of these systems was the devolution of control of the organisation from its previous political masters. In the case of (CMIS) there was also a perceived need, on the part of some senior managers within the organisation, for a system to monitor more closely the activities of the business. These involve the work of many, previously autonomous professionals, forced by changes in circumstances to allow much greater scrutiny of their activities.

The devolution of control necessitated the fast development of systems to enable survival. Financial systems were given priority and developed relatively quickly. These were perceived as satisfactory, primarily because they worked sufficiently well, to enable the organisation to survive financially. A personnel system was also developed. (CMIS) was the most difficult system to be attempted. (CMIS) could be considered as a customer database, plus a product database, the primary requirements as outputs, being the fast production of customer invoices, and provision of statistics to 'the centre' and to central government.

The development of (CMIS) followed the traditional pattern, the project team for the systems development consisted of a team seconded from one division of the organisation.

Development proceeded, many within the organisation were apparently consulted about desired outputs from the system. Expectations were high that an important information tool was to be provided, which would benefit all. A manager from outside the organisation was recruited to oversee all the systems. This manager did not survive long, and was replaced by a member of the original development team. At this point it seems that a decision to consolidate on the work done to date, and commence production was taken. For various reasons it also seems that only a part of the originally envisaged system was to be developed, what could be seen as the central core, the part most useful to the 'centre'.

Approval for this decision and proposed method of proceeding was sought from the top management of the 'centre' and the five divisions. They were introduced to the intricacies of the data model at a series of meetings. This approval was forthcoming, so production and implementation commenced. A steering group of the top management from each division, plus other interested parties, was set up to oversee further developments and make decisions as required.

Initial perception of a problem

The realisation that there were serious problems with the use of CMIS occurred when one division commissioned some simple information systems to assist in their management process. Many of the facilities to be included were already available as part of CMIS, yet it was not being used. Worse still its reputation, in this particular division was catastrophic. It had become what Pettigrew (1992) describes as an organisational scapegoat.

The primary operational users and providers of data to CMIS are the administrative workers in the divisions. They perceived the system to be difficult to use, unfriendly and not particularly useful. Their immediate managers do not regard the system highly, seeing it as a tool of the centre which has little relevance for themselves. They do little to ensure adequate use of the system, which has resulted in a build up of poor quality data. The poor quality data is itself a contributory factor in the limited benefits which are available from the system. The managers have little appreciation of the key organisational role the system could play.

On the other hand those at the centre seem unaware of the very real problems in using the system, particularly since there is little help available, and those in the divisions are overburdened with other responsibilities. The unfriendly interface became apparent, with the introduction of the 'Windows' environment as a standard for other applications. Administrative staff, some of whom may not be very comfortable with IT, have to move from a multi-coloured WIMP environment one minute, to old-fashioned green text on a dark screen, needing complex function key use, the next.

A turning point

The poor use and support of (CMIS) had a serious impact on its possible use for its major original aim - providing central management information and statistics. This led to the issue of a major directive from the controlling, steering body. The divisions were to start using a section of the system and complete data entry within a defined timescale. The directive caused some consternation within the divisions, evoked a number of strongly worded responses, but overall caused little in the way of action to improve system use.

In order to further inform thinking about the situation a project was undertaken which required intensive use of (CMIS). This was to provide a 'users' view of the system, to offer a perspective of it as an information system. The conclusion from this work was that (CMIS) was a perfectly adequate management information system, which potentially offered considerable benefits to the division. It was not a particularly easy system to use, and was not supported as well as it might be, but nonetheless could be made to work reasonably well, once the user was familiar with its complex interface.

The real problem ?

The real problem seemed to be system ownership. To those at the centre this is an essential tool to enable them to provide statistical data to a central government body. They are perceived as the owners of the system. The system is dependant on the data which is provided by those in the divisions, but it does little for them - they feel no ownership. More than this they feel hostility - that it is an imposition, which prevents them completing the work that they need to do.

Why has this problem arisen ?

One possible cause is the manner in which the IS professionals tried to involve divisional managers in the development stages of the system. The design methodology required user participation at a series of meetings. At these meetings the developers presented some of the system documentation to the assembled managers - seeking approval or signing-off, of particular stages. This attempt at facilitating participation was seen by the managers as disastrous - they did not understand the diagrams and documentation presented - were unable to comprehend the decisions required and reacted with hostility to the IS professionals, and the concept of the system.

Several of the managers tell the same story of this episode, being presented with complex wiring diagrams, and being expected to make decisions based on incomprehensible data. As managers in this organisation of course, they only admitted to the lack of comprehension later, and never to the IT professionals. They were culturally unable to say 'I don't understand - please explain?' Equally the IS professionals were apparently unable to understand or appreciate this lack of comprehension, and might well have been unable to explain meaningfully. Thus the vicious circle of negative perceptions began, the unintended consequence of a well meaning attempt at encouraging user participation and involvement.

It would appear that the inability of the IS professionals to communicate with the managers at this point is a classic 'culture gap' scenario - but there is more to the story than this. The steering group which was set up to oversee the continued system development and make decisions as required, was another attempt to improve communication, to overcome the 'culture gap', once more not very successful. Once technical items start to appear on the agenda, senior management attendance fell, soon those that attend, are those to whom the task is delegated as a chore, and a few who have a genuine interest. Yet the body is still a quasi-decision maker, and holder of some political power. Defining a suitable confederation of delegates to this group is the real challenge for the future. People who can act as a 'next friend' to the system, and facilitate its use, regardless of their hierarchical position. Taking this idea on board however, would probably be a culture-shock for some conservative central managers.

To sum up the situation:-

Divisional managers,
- do not understand system capabilities
- feel no system ownership

- feel threatened by central control
- see the system as a central control tool.

Central managers,

- always under pressure from national body
- do not understand divisional problems
- maintain central control while paying lip service to de-centralisation
- fail to appreciate divisions real need for IS.

Thus this is not a culture gap situation between IT professionals and business managers, but a conflict situation between central and divisional managers. The conflict occurs because of the differing goals of the sub-systems involved in the situation, both have needs which are crucial to the organisation, they pursue those which they perceive as most important to their own sub-system, failing to take the holistic organisational perspective. The CMIS system caught in the midst of this conflict, suffers, a victim of a political culture, where being seen to do the politically correct, is assumed to be sufficient for effectiveness.

This is why it is more of a trap than a gap - the organisational culture, a memo-culture, mediates against realistic communication, and any real understanding of the other side's viewpoint - it traps both sides in a situation that cries out for metanoia.

It is a trap also because it is not seen, this way of behaving is very much part of the adversarial, confrontational and political culture. The roles, values and norms are those of the battlefield. Fuenmayor (1995) quoting Geoffrey Vickers has some ideas relevant to this form of trap:-

> a trap is only a trap for creatures who cannot solve the problems that it sets... the nature of the trap is a function of the nature of the trapped...we the trapped, tend to take our own state of mind for granted...which is partly why we are trapped.

thus the actors within the organisation create their own trap, which surrounds, envelops and controls because it is not observed, not challenged, until it is too late and doom is inevitable. It is a powerful trap because it the trap of 'the way we do things round here' reinforced by fear of the unknown and reluctance to change.

What of the organisation in the case study, is it trapped in a downward spiral moving to its doom? Perhaps not, hope is offered by Schein(1981), who suggests that the norms in an organisation have in fact two components. These are pivotal norms, observance of which are mandatory for continued membership of the organisation, but also peripheral norms observance of which is desirable for conformance. Those who accept pivotal norms but challenge peripheral ones he describes as 'creative individualists', these people are capable of creativity on behalf of the organisation, this creativity may lead to innovation in a number of ways within the organisation, and a movement away from the downward spiral to doom. A dialogue could move the situation forward, on dialogue the words from Senge (1990) seem appropriate:-

> We are not trying to win in dialogue. We all win if we are doing it right. In dialogue, individuals gain insights that simply could not be achieved individually. A new kind of mind comes into being which is based on the development of common meaning

The final words though, must echo those of Peter Checkland(1981), the situation here, as in so many problem situations, is the search for that which is both *'systemically desirable and culturally feasible'*, the challenge perhaps, is to persuade the management side to learn more about what is systemically desirable, the IT side to understand more of what is culturally feasible, and both sides that this can be a shared vision.

References

Ackoff Russell L. (1967), Management Misinformation Systems, *Management Science*, Dec. 1967.

Checkland P.B. (1981), *Systems Thinking, Systems Practice*, Chichester UK, Wiley.

Forrester, Jay, 1994, *Policies, Decisions, and Information Sources for Modelling, in Modelling for learning Organizations*, ed. Morecroft, J.D.W., & Sterman, J.D., Portland Oregon, Productiviy Press.

Forrester, Jay, 1971, Counter Intuitive Behaviour of Social Systems, *Technology Review 73*, No.3:52-68.

Fuenmayor, R., 1995, The will to systems, in *Critical Issues in Systems Theory and Practice*, ed. Ellis K., Plenum.

Galliers, R., 1991, Strategic information systems planning : myths, reality and guidelines for successful implementation, *European Journal of Information Systems*, 1, 1, 55-64.

Grindley, K., ed., 1992,Culture Gap, *The Price Warehouse IT Review of 1991/1992, London.*

Hammer, M., 1990, Reengineering Work: Don't Automate Obliterate, *Harvard Business Review*, July-August 1990.

Hammer, M. & Champy, J., 1993, *Reengineering the Corporation*, London, Nicholas Brealey.

Jackson, M. C., 1993, Beyond the fads:systems thinking for managers, *University of Hull working paper no.3.*

Kim, D. H., & Senge, P. M., 1994, Putting Systems Thinking into practice, *Systems Dynamics Review*, 10, 2-3.

Madnick, S., 1988, *The Strategic Use of Information technology*, New York, Oxford University Press.

Ormerod, R., 1994, Putting Soft OR Methods to work: Information Systems Strategy Development at Sainsbury's, *Warwick Business School research paper.*

Pettigrew, A., 1987, Context and action in the transformation of the firm, *Journal of Management Studies*, 24-6.

Porter M. E. & Millar V. E., (1985), How information gives you competitive advantage, *Harvard Business Review* 63(4).

Senge P. M. 1990, *The Fifth Discipline*, London, Century.

Ward, J.,Griffiths, P.,& Whitmore P., (1990), *Strategic Planning for Information systems*, Chichester, Wiley.

Venkatraman, N., 1991, *IT-Induced Business Reconfiguration, in The Corporation of the 1990's*, ed. Scott Morton, M. S., New York, Oxford University Press.

PART II
ORGANISATIONAL
PERSPECTIVE

4 Organisational Analysis in Computer Science

ROB KLING
DEPARTMENT OF INFORMATION & COMPUTER SCIENCE AND CENTER
FOR RESEARCH ON INFORMATION TECHNOLOGY AND
ORGANIZATIONS
UNIVERSITY OF CALIFORNIA AT IRVINE,
USA

Abstract

Computer Science is hard pressed in the US to show broad utility to help justify billion dollar research programs and the value of educating well over 40,000 Bachelor of Science and Master of Science specialists annually in the U.S. The Computer Science and Telecommunications Board of the U.S. National Research Council has recently issued a report, "Computing the Future (Hartmanis and Lin, 1992)" which sets a new agenda for Computer Science. The report recommends that Computer Scientists broaden their conceptions of the discipline to include computing applications and domains to help understand them. This short paper argues that many Computer Science graduates need some skills in analysing human organisations to help develop appropriate systems requirements since they are trying to develop high performance computing applications that effectively support higher performance human organisations. It is time for academic Computer Science to embrace organisational analysis (the field of Organisational Informatics) as a key area of research and instruction.

Introduction

Computer Science is being pressed on two sides to show broad utility for substantial research and educational support. For example, the High Performance Computing Act will provide almost two billion dollars for research and advanced development. Its advocates justified it with arguments that specific technologies, such as parallel computing and wideband nets, are necessary for social and economic development. In the US, Computer Science academic programs award well over 30,000

Bachelor of Science (BS) and almost 10,000 Master of Science (MS) degrees annually. Some of these students enter PhD programs and many work on projects which emphasise mathematical Computer Science. But many of these graduates also take computing jobs for which they are inadequately educated, such as helping to develop high performance computing applications to improve the performance of human organisations. These dual pressures challenge leading Computer Scientists to broaden their conceptions of the discipline to include an understanding of key application domains, including computational science and commercial information systems. An important report that develops this line of analysis, "Computing the Future" (CTF) (Hartmanis and Lin, 1992), was recently issued by the Computer Science and Telecommunications Board of the U.S. National Research Council.

CTF is a welcome report that argues that academic Computer Scientists must acknowledge the driving forces behind the substantial Federal research support for the discipline. The explosive growth of computing and demand for CS in the last decade has been driven by a diverse array of applications and new modes of computing in diverse social settings. CTF takes a strong and useful position in encouraging all Computer Scientists to broaden our conceptions of the discipline and to examine computing in the context of interesting applications.

CTF's authors encourage Computer Scientists to envision new technologies in the social contexts in which they will be used. They identify numerous examples of computer applications in earth science, computational biology, medical care, electronic libraries and commercial computing that can provide significant value to people and their organisations. These assessments rest on concise and tacit analyses of the likely design, implementation within organisations, and uses of these technologies. For example, CTF's stories of improved computational support for modelling are based on rational models of organisational behaviour. They assume that professionals, scientists, and policy-makers use models to help improve their decisions. But what if organisations behave differently when they use models? For example suppose policy makers use models to help rationalise and legitimise decisions which are made without actual reference to the models?

One cannot discriminate between these divergent roles of human organisations based upon the intentions of researchers and system designers. The report tacitly requires that the CS community develop reliable knowledge, based on systematic research, to support effective analysis of the likely designs and uses of computerised systems. CTF tacitly requires an ability to teach such skills to CS practitioners and students. Without a disciplined skill in analysing human organisations, Computer Scientists' claims about the usability and social value of specific technologies is mere opinion, and bears a significant risk of being misleading. Further, Computer Scientists who do not have refined social analytical skills sometimes conceive and promote technologies that are far less useful or more costly than they claim. Effective CS practitioners who "compute for the future" in organisations need some refined skills in organisational analysis to understand appropriate systems requirements and the conditions that transform high performance computing into high performance human organisations. Since CTF does not spell out these tacit implications, I'd like to explain them here.

Broadening computer science: from computability to usability

The usability of systems and software is a key theme in the history of CS. We must develop theoretical foundations for the discipline that give the deepest insights in to what makes systems usable for various people, groups and organisations. Traditional computer scientists commonly refer to mathematics as the theoretical foundations of CS. However, mathematical formulations give us limited insights into understanding why and when some computer systems are more usable than others.

Certain applications, such as supercomputing and computational science are evolutionary extensions of traditional scientific computation, despite their new direction with rich graphical front ends for visualising enormous mounds of data. But other, newer modes of computing, such as networking and microcomputing, change the distribution of applications. While they support traditional numerical computation, albeit in newer formats such as spreadsheets, they have also expanded the diversity of non-numerical computations. They make digitally represented text and graphics accessible to tens of millions of people.

These technological advances are not inconsistent with mathematical foundations in CS, such as Turing machine formulations. But the value of these formats for computation is not well conceptualised by the foundational mathematical models of computation. For example, text editing could be conceptualised as a mathematical function that transforms an initial text and a vector of incremental alterations into a revised text. Text formatting can be conceptualised as a complex function mapping text strings into spatial arrays. These kinds of formulations don't help us grasp why many people find "what you see is what you get" editors as much more intuitively appealing than a system that links line editors, command-driven formatting languages, and text compilers in series.

Nor do our foundational mathematical models provide useful ways of conceptualising some key advances in even more traditional elements of computer systems such as operating systems and database systems. For example, certain mathematical models underlie the major families of database systems. But one can't rely on mathematics alone to assess how well networks, relations, or object-entities serve as representations for the data stored in an airline reservation system. While mathematical analysis can help optimise the efficiency of disk space in storing the data, they can't do much to help airlines understand the kinds of services that will make such systems most useful for reservationists, travel agents and even individual travellers. An airline reservation system in use is not simply a closed technical system. It is an open socio-technical system (Hewitt, 1986; Kling, 1992). Mathematical analysis can play a central role in some areas of CS, and an important role in many areas. But we cannot understand important aspects of usability if we limit ourselves to mathematical theories.

The growing emphasis of usability is one of the most dominant of the diverse trends in computing. The usability tradition has deep roots in CS, and has influenced the design of programming languages and operating systems for over 25 years. Specific topics in each of these areas also rest on mathematical analysis which Computer Scientists could point to as "the foundations" of the respective subdisciplines. But Computer Scientists envision many key advances as design conceptions rather than as mathematical theories. For example, integrated programming environments ease software development. But their conception and popularity is not been based on deeper formal foundations for programming languages. However, the growth of non-numerical applications for diverse professionals, including text processing, electronic mail, graphics, and multimedia should place a premium on making computer systems relatively simple to use. Human Computer Interaction (HCI) is now considered a core subdiscipline of CS.

The integration of HCI into the core of CS requires us to expand our conception of the theoretical foundations of the discipline. While every computational interface is reducible to a Turing computation, the foundational mathematical models of CS do not (and could not) provide a sound theoretical basis for understanding why some interfaces are more effective for some groups of people than others. The theoretical foundations of effective computer interfaces must rest on sound theories of human behaviour and their empirical manifestations (cf. Ehn, 1991, Grudin, 1989).

Interfaces also involve capabilities beyond the primary information processing features of a technology. They entail ways in which people learn about systems and ways to manage the diverse data sets that routinely arise in using many computerised systems (Kling, 1992). Understanding the diversity and character of these interfaces, that are required to make many systems usable, rests in an understanding the way that people and groups organise their work and expertise with computing. Appropriate theories of the diverse interfaces that render many computer systems truly useful must rest, in part, on theories of work and organisation. There is a growing realisation, as networks tie users together at a rapidly rising rate, that usability cannot generally be determined without our considering how computer systems are shaped by and also alter interdependencies in groups and organisations. The newly-formed subdiscipline of Computer Supported Co-operative Work and newly-coined terms "groupware" and "co-ordination theory" are responses to this realisation (Greif, 1988; Galegher, Kraut and Egido, 1990).

Broadening computer science: from high performance computing to high performance organisations

The arguments of CTF go beyond a focus on usable interface designs to claims that computerised systems will improve the performance of organisations. The report argues that the US should invest close to a billion dollars a year in CS research because of the resulting economic and social gains. These are important claims, to which critics can seek systematic evidence. For example, one can investigate the claim that 20 years of major computing R&D and corporate investment in the US has helped provide proportionate economic and social value.

CTF is filled with numerous examples where computer-based systems provided value to people and organisations. The tough question is whether the overall productive value of these investments is worth the overall acquisition and operation costs. While it is conventional wisdom that computerisation must improve productivity, a few researchers began to see systemic possibilities of counter-productive computerisation in the early 1980s (King and Kraemer, 1981). In the last few years economists have found it hard to give unambiguously affirmative answers to this question. The issue has been termed "The Productivity Paradox," based on a comment attributed to Nobel laureate Robert Solow who remarked that "computers are showing up everywhere except in the [productivity] statistics (Dunlop and Kling, 1991a)".

Economists are still studying the conditions under which computerisation contributes to organisational productivity, and how to measure it [1]. But even if computerisation proves to be a productive investment, in the net, in most economic sectors, there is good reason to believe that many organisations get much less value from their computing investments than they could and should.

There is no automatic link between computerisation and improved productivity. While many computer systems have been usable and useful, productivity gains require that their value exceed all of their costs.

There are numerous potential slips in translating high performance computing into cost-effective improvements in organisational performance. Some technologies are superb for well-trained experts, but are difficult for less experienced people or "casual users." Many technologies, such as networks and mail systems, often require extensive technical support, thus adding hidden costs (Kling, 1992).

Further, a significant body of empirical research shows that the social processes by which computer systems are introduced and organised makes a substantial difference in their value to people, groups and organisations (Lucas, 1981; Kraemer, et. al. 1985; Orlikowski, 1992). Most seriously, not all presumably appropriate computer applications fit a person or group's work practices. While they may make sense in a simplified world, they can actually complicate or misdirect real work.

Group calendars are but one example of systems that can sound useful, but are often useless because they impose burdensome record keeping demands (Grudin, 1989). In contrast, electronic mail is one of the most popular applications in office support systems, even when other capabilities, like group calendars, are ignored (Bullen and Bennett, 1991). However, senders are most likely to share information with others when the system helps provide social feedback about the value of their efforts or they have special incentives (Sproull and Kiesler, 1991; Orlikowski, 1992). Careful attention to the social arrangements or work can help Computer Scientists improve some systems designs, or also appreciate which applications may not be effective unless work arrangements are changed when the system is introduced.

The uses and social value of most computerised systems can not be effectively ascertained from precise statements of their basic design principles and social purposes. They must be analysed within the social contexts in which they will be used. Effective social analyses go beyond accounting for formal tasks and purposes to include informal social behaviour, available resources, and the interdependencies between key groups (Cotterman and Senn, 1992).

Many of the BS and MS graduates of CS departments find employment on projects where improved computing should enhance the performance of specific organisations or industries. Unfortunately, few of these CS graduates have developed an adequate conceptual basis for understanding when information systems will actually improve organisational performance. Consequently, many of them are prone to recommend systems-based solutions whose structure or implementation within organisations would be problematic.

Organisational informatics

Organisational Informatics denotes a field which studies the development and use of computerised information systems and communication systems in organisations. It includes studies of their conception, design, effective implementation within organisations, maintenance, use, organisational value, conditions that foster risks of failures, and their effects for people and an organisation's clients. It is an intellectually rich and practical research area.

Organisational Informatics is a relatively new label. In Europe, the term Informatics is the name of many academic departments which combine both CS and Information Systems. In North America, Business Schools are the primary institutional home of Information Systems research and teaching. But this location is a mixed blessing. It brings IS research closer to organisational studies. But the institutional imperatives of business schools lead IS researchers to emphasise the development and use of systems in a narrow range of organisations -- businesses generally, and often service industry firms. It excludes information systems in important social sectors such as health care, military operations, air-traffic control, libraries, home uses, and so on. And Information Systems research tries to avoid messy issues which many practising Computer Scientists encounter: developing requirements for effective systems and mitigating the major risks to people and organisations who depend upon them.

The emerging field of Organisational Informatics builds upon research conducted under rubrics like Information Systems and Information Engineering. But it is more wide ranging than either of these fields are in practice[2].

Organisational informatics research

In the last 20 years a loosely organised community of some dozens of researchers have produced a notable body of systematic scientific research in Organisational Informatics. These studies examine a variety of topics, including:

- How system designers translate people's preferences into requirements;
- The functioning of software development teams in practice;
- The conditions that foster and impede the implementation of computerised systems within organisations;
- The ways that computerised systems simplify or complicate co-ordination within and between organisations;
- How people and organisations use systems in practice;
- The roles of computerised systems in altering work, group communication, power relationships, and organisational practices.

Researchers have extensively studied some of these topics, such as computerisation and changing work, appear in synoptic review articles (Kling and Dunlop, in press). In contrast, researchers have recently begun to examine other topics, such software design (Winograd and Flores, 1986; Kyng and Greenbaum, 1991), and have recently begun to use careful empirical methods (e.g. Suchman, 1983; Bentley, et. al. 1992; Fish, et. al., 1993). I cannot summarise the key theories and rich findings of these diverse topics in a few paragraphs. But I would like to comment upon a few key aspects of this body of research.

Computer systems use in social worlds

Many studies contrast actual patterns of systems design, implementation, use or impacts with predictions made by Computer Scientists and professional commentators. A remarkable fraction of these accounts are infused with a hyper-rational and under- socialised view of people, computer systems, organisations and social life in general. Computer Scientists found that rule driven conceptions to be powerful ways to abstract domains like compilers. But many Computer Scientists extend them to be a tacit organising frame for understanding whole computer systems, their developers, their users and others who live and work with them. Organisations are portrayed as generally co-operative systems with relatively simple and clear goals. Computer systems are portrayed as generally coherent and adequate for the tasks for which people use them. People are portrayed as generally obedient and co-operative participants in a highly structured system with numerous tacit rules to be obeyed, such as doing their jobs as they are formally described. Using data that is contained in computer systems, and treating it as information or knowledge, is a key element of these accounts. Further, computer systems are portrayed as powerful, and often central, agents of organisational change.

This Systems Rationalist perspective infuses many accounts of computer systems design, development, and use in diverse application domains, including CASE tools, instructional computing, models in support of public policy assessments, expert systems, groupware, supercomputing, and network communications (Kling, 1980; Kling, Scherson and Allen, 1992).

All conceptual perspectives are limited and distort "reality." When Organisational Informatics researchers systematically examine the design practices in particular organisations, how specific groups develop computer systems, or how various people and groups use computerised systems, they find an enormous range of fascinating and important human behaviour which lies outside the predictive frame of Systems Rationalism. Sometimes these behaviours are relatively minor in overall importance. But in many cases they are so significant as to lead Organisational Informatics researchers to radically reconceptualise the processes which shape and are shaped by computerisation.

There are several alternative frames for reconceptualising computerisation as alternatives to Systems Rationalism. The alternatives reflect, in part, the paradigmatic diversity of the social sciences. But all of these reconceptions situate computer systems and organisations in richer social contexts and with more complex and multivalent social relations than does systems rationalism. Two different kinds of observations help anchor these abstractions.

Those who wish to understand the dynamics of model usage in public agencies must appreciate the institutional relationships which influence the organisation's behaviour. For example, to understand economic forecasting by the US Congress and the U.S. Executive branch's Office of Management and Budget, one must appreciate the institutional relations between them. They are not well described by Systems Rationalist conceptions because they were designed to continually differ with each other in their perspectives and preferred policies. That is one meaning of "checks and balances" in the fundamental design of the US Federal Government. My colleagues, Ken Kraemer and John King, titled their book about Federal economic modelling, Data Wars (Kraemer, et. al., 1985). Even this title doesn't make much sense within a Systems Rationalist framework.

Modelling can be a form of intellectual exploration. It can also be a medium of communication, negotiation, and persuasion. The social relationships between modellers, people who use them and diverse actors in Federal policymaking made these socially mediated roles of models sometimes most important. In these situations, an alternative view of organisations as coalitions of interest groups was a more appropriate conceptualisation. And within this coalitional view of organisations, a conception of econometric models as persuasion support systems rather than as decision support systems sometimes is most appropriate. Organisational Informatics researchers found that political views of organisations and systems developments within them apply to many private organisations as well as to explicitly political public agencies.

Another major idea to emerge from the broad body of Organisational Informatics research is that the social patterns which characterise the design, development, uses and consequences of computerised systems are dependent on the particular ecology of social relationships between participants. This idea may be summarised by saying that the processes and consequences of computerisation are "context dependent." In practice, this means that the analyst must be careful in generalising from one organisational setting to another. While data wars might characterise econometric modelling on Capitol Hill, we do not conclude that all computer modelling should be interpreted as persuasion support systems. In some settings, models are used to explore the effects of policy alternatives without immediate regard for their support as media for communication, negotiation or persuasion. At other times, the same model might be used (or abused with cooked data) as a medium of persuasion. The brief accounts of models for global warming in CTF fit a Systems Rationalist account. Their uses might appear much less "scientific" if they were studied within the actual policy processes within which they are typically used.

Computing in a web of technological and social dependencies: the role of infrastructure

Another key feature of computerised systems is the technological and organisational infrastructure required to support their effective use (Kling and Scacchi, 1982; Kling, 1987; Kling, 1992). The information processing models of computerised systems focus on the "surface structures," such as information flows within a system. For example, one can compare the information processing capabilities of computerised modelling systems in terms of the complexity and variety of computations that they support, the richness of their graphical displays, and so on. Text processing systems can be similarly compared by contrasting their capabilities for handling footnotes, graphics, fine grained text placement, custom dictionaries and so on. From an information processing point of view, system A is usually better than system B if it offers many more capabilities than system B. Information processing conceptions have also fuelled much of the talk about high performance computing. It is common to talk about massively parallel computing in terms of the scale and unit cost of computation (Kling,

Scherson, and Allen 1992), and the discussions of networking in terms of the wide data bandwidths that new technologies offer.

If we ask how these technologies improve organisational performance, then we have to ask how they can be made usable to diverse groups. The most powerful modelling system may be of limited utility if it requires sophisticated programming skills to create and modify every data transformation. Alternatively, such a package can be made more widely useful by having the modelling efforts managed by a programming group which provides added value for added cost.

Few people are capable or interested in primarily using "raw computing" for their work. The diverse array of "productivity software" -- such as text processing, presentation graphics, spreadsheets, databases and so on gain their value when they can be provided and maintained in a way that matches the skills and available time of people who will use them. Both skill and time are scarce resources in most organisations. Skilled time is especially expensive.

Similarly, the organisational value of digital libraries can't be adequately conceptualised in terms of simple data-centric measures, like the number of gigabytes of available files. The ease of people accessing useful documents is much more pertinent, although much less frequently discussed today.

In each of these cases, the support systems for the focal computing system is integral to the effective operation of the technology. Infrastructure refers to the set of human and organisational resources that help make it simpler and faster for skilled people to use computerised systems. Infrastructure should be part of the conceptualisation. Often the support systems for a computing can involve several different organisations, including hardware and software vendors, telecommunication support groups, divisional systems groups, and local experts (Kling, 1992). It can be organisationally very complex and unresponsive in some cases and organisationally simpler and more effective in others. In any case, the infrastructure for systems support can't be ignored when one is interested in improving organisational performance.

Repercussions for systems design

Even when computerised systems are used as media of intellectual exploration, Organisational Informatics researchers find that social relationships influence the ways that people use computerised systems. Christine Bullen and John Bennett (1991) studied 25 organisations that used groupware with diverse modules such as databases, group calendars, text annotating facilities and electronic mail. They found that the electronic mail modules were almost universally valued, while other system facilities were often unused.

In a recent study, Sharyn Ladner and Hope Tillman examined the use of the Internet by university and corporate librarians. While many of them found data access through databases and file transfer to be important services, they also reported that electronic mail was perhaps the most critical Internet feature for them.

> The participants in our study tell us something that we may have forgotten in our infatuation with the new forms of information made available through the Internet. And that is their need for community. To be sure, our respondents use the Internet to obtain information not available in any other format, to access databases ... that provide new efficiencies in their work, new ways of working. But their primary use is for communication. Special librarians tend to be isolated in the workplace -- the only one in their subject speciality (in the case of academe), or the only librarian in their organisation (in the case of a corporate library). Time and time again our respondents expressed this need to talk to someone -- to learn what is going on in their profession, to bounce ideas off others, to obtain information from people, not machines. There are tremendous implications from the Internet technology in community formation -- the Internet may indeed provide a way to increase community among scholars, including librarians. The danger we face at this juncture in time, as we attach library resources to the Internet, is to focus all of our energies on the machine-based resources at the expense of our human-based resources, i.e., ourselves. - (Ladner and Tillman, 1992).

In these studies, Organisational Informatics researchers have developed a socially rich view of work with and around computing, of computing within a social world.

These studies have strong repercussions for the design of software. A good designer cannot assume that the majority of effort should go into the "computational centerpiece" of a system, while devoting minor efforts to supporting communication facilities. One of my colleagues designed a modelling system for managers in a major telephone company, after completing an extensive requirements analysis. However, as an afterthought, he added a simple mail system in a few days work. He was surprised to find that the people who used these systems regularly used his crude electronic mail system, while they often ignored interesting modelling capabilities. Such balances of attention also have significant repercussions. Many people need good mail systems, not just crude ones: systems which include facile editors, ease in exporting and importing files, and effective mail management (Kling and Covi, 1993).

Assessing people's preferences for systems' designs is an exercise in social inquiry. While rapid prototyping may help improve designs for some systems, it is less readily applicable to systems which are used by diverse groups at numerous locations. Computer scientists are beginning to develop more reliable methods of social inquiry to better understand which systems designs will be most useful (Bentley, et. al. 1992; Kyng and Greenbaum, 1991). It is particularly helpful to organise system designs that help minimise the complexity and cost of its infrastructure (Kling, 1992).

Fish and his colleagues (1993) recently reported the way that the explicit use of social theory helped them design more effective group meeting systems. Unfortunately, these newer methods are rarely taught to CS students. When computer specialists build an imbalanced system, it should not be a surprise when the resulting organisational value of their efforts is very suboptimal.

System security and reliability

In a simplified engineering model of computing, the reliability of products is assured through extensive testing in a development lab. The social world of technology use not perceived as shaping the reliability of systems, except through irascible human factors, such as "operator errors." An interesting and tragic illustration of the limitations of this view can be found in some recent studies of the causes of death and maiming by an electron accelerator which was designed to help cure cancer, the Therac-25 (Jacky, 1991, Leveson and Turner, 1993).

The Therac-25 was designed and marketed in the mid 1980s by a Canadian firm, Atomic Energy of Canada Limited (AECL), as an advanced medical technology. It featured complete software control over all major functions (supported by a DEC PDP-11), among other innovations. Previous machines included electro- mechanical interlocks to raise and lower radiation shields. Several thousand people were effectively treated with the Therac- 25 each year. However, between 1985 and 1987 there were six known accidents in which several people died in the US. Other were seriously maimed or injured [3].

Both studies concur that there were subtle but important flaws in the design of the Therac-25's software and hardware. AECL's engineers tried to patch the existing hardware and (finally) software when they learned of some of the mishaps. But they treated each fix as the final repair.

Both studies show how the continuing series of mishaps was exacerbated by diverse organisational arrangements. Jacky claims that pressures for speedy work by radiological technicians coupled with an interface design that did not enhance important error messages was one of many causes of the accidents. Leveson and Turner differ in downplaying the working conditions of the Therac-25's operators and emphasise the flawed social system for communicating the seriousness of problems to Federal regulators and other hospitals. Both studies observe that it is unlikely for the best of companies to develop perfect error-free systems without high quality feedback from users. Their recommendations differ: Jacky discusses the licensing of system developers and the regulation of computerised medical systems to improve minimal standards of safety. Leveson and Turner propose extensive education and training of software engineers and more effective communication between manufacturers and their customers.

However, both studies indicate that an understanding of the safety of computer systems must go beyond the laboratory and extend into the organisational settings where it is used. In the case of the Therac-25, it required understanding a complex web of Interorganisational relationships, as well as the technical design and operation of the equipment. Nancy Leveson (1992) points out that most major disasters technological disasters in the last 20 years "involved serious organisational and management deficiencies." Hughes, Randall and Shapiro (1992:119) observe that British no civil collision in UK air space has been attributed to air traffic control failures. But their Mediator control system was failing regularly and had no backup during the period that they studied it. They observe that the reliability of the British air traffic control system resides in totality of the relevant social and technical systems, rather than in a single component.

The need for this kind of organisational understanding is unfortunately slighted in the CS academic world today. CTF discusses only those aspects of computer system reliability which are amenable to understanding through laboratory-like studies (Hartmanis and Lin, 1992:110-111). But cases of safety critical systems, like the Therac-25 and British Air Traffic Control, indicate why some Computer Scientists must be willing to undertake (and teach) organisational analysis.

Worldviews and surprises about computerisation

These few paragraphs barely sketch the highlights of a fertile and significant body of research about computer systems in use. Perhaps the most important simplification for traditional computer scientists is to appreciate how people and their organisations are situated in a social world and consequently compute within a social world. People act in relationship to others in various ways and concerns of belonging, status, resources, and power are often central. The web of people's relationships extend beyond various formally defined group and organisational boundaries (Kling and Scacchi, 1982; Kling, 1987; Kling, 1992). People construct their worlds, including the meanings and uses of information technologies, through their social interactions.

This view is, of course, not new to social scientists. On the other hand, there is no specific body of social theory which can easily be specialised for "the case of computing," and swiftly produce good theories for Organisational Informatics as trivial deductions. The best research in Organisational Informatics draws upon diverse theoretical and methodological approaches within the social sciences with a strong effort to select those which best explain diverse aspects of computerisation.

Organisational informatics within computer science

CTF places dual responsibilities on Computer Scientists. One responsibility is to produce a significant body of applicable research. The other responsibility is to educate a significant fraction of CS students to be more effective in conceiving and implementing systems that will enhance organisational performance. It may be possible to organise research and instruction so as to decouple these responsibilities. For example, molecular biologists play only a small role in training doctors. However, CS departments act like an integrated Medical school and Biology department. They are the primary academic locations for training degreed computing specialists, and they conduct a diverse array of less applicable and more applicable research. In practice, the research interests of CS faculty shape the range of topics taught in CS departments, especially the 150 PhD granting departments. CS curricula mirror major areas of CS research and the topics which CS faculty understand through their own educations and subsequent research. As a consequence, CS courses are likely to avoid important CS topics which appear a bit foreign to the instructor.

An interesting example of this coupling can be illustrated by CTF, in a brief description of public-key encryption systems and digital signatures (Hartmanis and Lin, 1992:27). In the simple example, Bob and Alice can send messages reliably if each maintains a secret key. Nothing is said about the social complications of actually keeping keys secret. The practical problems are similar to those of managing passwords, although some operational details differ because the 100 digit keys may be stored on media like magstripe cards rather than paper. In real organisations, people lose or forget their password and can lose the media which store their keys. Also, some passwords can be shared by a group of with shifting membership, and the "secret key" can readily become semi-public. The main point is that the management of keys is a critical element of cryptographic security in practice. But Computer Scientists are prone to teach courses on cryptography as exercises in applied mathematics, such as number theory and Galois theory, and to skirt the vexing practical problems of making encryption a practical organisational activity.

Today, most of the 40,000 people who obtain BS and MS degrees in CS each year in the U.S. have no opportunities for systematic exposure to reliable knowledge about the best design strategies, common uses, effective implementation, and assessments of value of computing in a social world (Lewis, 1989). Yet a substantial fraction of these students go on to work for organisations attempting to produce or maintain systems that improve organisational performance without a good conceptual basis for their work. Consequently, many of them develop systems that underperform in organisational terms even when they are technically refined. They also recommend ineffective implementation procedures and are sometimes even counterproductive.

One defensible alternative to my position is that CS departments should not take on any form of organisational analysis. They should aggressively take a role akin to Biology departments rather than taking on any instructional or research roles like Medical schools. To be sincere, this position requires a high level of restraint by academic Computer Scientists. First and foremost, they should cease from talking about the uses, value or even problems of computerised systems that would be used in any organisational setting. Research proposals would be mute about any conceivable application of research results. Further, they should make effective efforts to insure that anyone who employs their graduates should be aware that they may have no special skills in understanding organisational computing. It would take an aggressive "truth in advertising" campaign to help make it clear that Computer Scientists have no effective methods for understanding computerisation in the social world. Further, Computer Scientists would forsake their commitments to subfields like software engineering which tacitly deals with ways to support teams of systems developers to work effectively (Curtis, et. al. 1988). Computer Scientists, in this view, would remove themselves from addressing organisational and human behaviour, in the same way that molecular biologists are removed from professionally commenting on the practices of cardiologists and obstetricians. CTF argues that this view would be self-defeating. But it would be internally consistent and have a distinctive integrity.

In contrast, CS faculty are often reluctant to wholly embrace Organisational Informatics. But some CS subfields, such as software engineering, depend upon organisational analysis (Curtis, et. al., 1988). Further, CS faculty do little to advertise the distinctive limitations in the analytical skills of our programs' graduates. Part of the dilemma develops because many CS faculty are ambivalent about systematic studies of human behaviour. Applied mathematics and other modes of inquiry which seem to yield concise, crisp and concrete results are often the most cherished. As a consequence, those who conduct behaviourally oriented research in CS departments are often inappropriately marginalised. Their students and the discipline suffers as a result.

Between 1986 and 1989, the total number of BS and MS CS degrees awarded annually in the US declined from about 50,000 to approximately 40,000. The number of students majoring in CS rapidly declined at a time when computerisation was becoming widespread in many fields. A significant fraction of the decline can be attributed to many students finding CS programs insular and indifferent to many exciting forms of computerisation. The decline of military R&D in the U.S. can amplify these trends or stimulate a more cosmopolitan view in CS departments. The decline in military R&D is shifting the job market for new CS graduates towards a markedly more civilian orientation. This shift, along with the trend towards computing distributed into diverse work groups, is leading to more job opportunities for people with CS education who know Organisational Informatics.

The situation of CS departments has some parallels with Statistics departments. Statistics are widely used and taught in many academic disciplines. But Statistics departments have often maintained a monkish isolation from "applications." Consequently, the application of statistics thrives while Statistics departments have few students and modest resources. Might the status of Statistics indicate a future possibility for an insular approach to CS?

The best Organisational Informatics research in North America is conducted by faculty in the Information Systems departments in business schools and by scattered social scientists (cf. Boland and Hirschheim, 1987; Galegher, Kraut and Egido, 1990; Cotterman and Senn, 1992; Sproull and Kiesler, 1991). But Computer Scientists cannot effectively delegate the research and teaching of Organisational Informatics to business Schools or social science departments.

Like Computer Scientists, faculty in these other disciplines prefer to focus on their own self-defined issues. Computer Scientists are much more likely to ask questions with attention to fine grained technological nuances that influence designs. For example, the professional discussions of computer risks have been best developed through activities sponsored by the ACM's Special Interest Group on Software (SIGSOFT). They are outside the purview of business school faculty and, at best, only a few social scientists are interested in them. Generally, technology plays a minor role in social science theorising. And when social scientists study technologies, they see a world of possibilities: energy technologies, transportation technologies, communication technologies (including television), medicinal drugs and devices, and so on. They see little reason to give computer-related information technologies a privileged role within this cornucopia. As a consequence, the few social scientists who take a keen interest in studying computerisation are unfortunately placed in marginal positions within their own disciplines. Often they must link their studies to mainstream concerns as defined by the tastemakers of their own fields, and the resulting publications appear irrelevant to Computer Scientists.

Further, faculty in these other disciplines are not organised to effectively teach tens of thousands of CS students, students who are steeped in technology and usually very naive about organisations, about systems development and use in organisations. In North America there is no well developed institutional arrangement for educating students who can effectively take leadership roles in conceptualising and developing complex organisational computing projects (Lewis, 1989).

CTF is permeated with interesting claims about the social value of recent and emerging computer-based technologies. While many of these observations should rest on an empirically grounded scientific footing, Computer Scientists have deprived themselves of access to such research. For example, the discussion of systems risks in the ACM rests on a large and varied collection of examples and anecdotes. But there is no significant research program to help better understand the conditions under which organisations are more likely to develop systems using the best risk-reducing practices. There is an interesting body of professional lore, but little scholarship to ground it (See Appendix).

Computer Scientists have virtually no scholarship to utilise in understanding when high performance networks, like the National Research and Education Network, will catalyse social value proportional to their costs. Consequently, many of the "obvious" claims about the value of various computing technologies that we Computer Scientists make are more akin to the lore of home remedies for curing illness. Some are valid, others are unfounded speculation. More seriously, the theoretical bases for recommending home medical remedies and new computer technologies can not advance without having sound research programs.

What is needed

CTF sets the stage for developing Organisational Informatics as a strong subfield within Computer Science. CTF bases the expansion of the discipline on a rich array of applications in which many of the effective technologies must be conceived in relationship to plausible uses in order provide attractive social value for multi-billion dollar public investments.

The CS community needs an institutionalised research capability to produce a reliable body of knowledge about the usability and value of computerised systems and the conditions under which computer systems improve organisational performance. In Western Europe there are research projects about Organisational Informatics in a few Computer Science departments and research funding through the EEC's Espirit program (Bubenko, 1992; Iivari, 1991; Kyng and Greenbaum, 1991). These new research and instructional programs in Western Europe give Organisational Informatics a significantly more effective place in CS education and research than it now has in North America.

The CS community in the U.S. has 30 years of experience in institutionalising research programs, especially through the Defense Advanced Research Projects Agency and the National Science Foundation (NSF). There are many approaches, including establishing national centers, supporting individual investigator research grants, supporting short institutes to help train new investigators and supporting research workshops for ongoing research. All such programs aim to develop and sustain research fields with a combination of direct research funds, the education of future researchers, and the development of research infrastructure. They are all multimillion dollar efforts. Today, NSF devotes about $125K annually to Organisational Informatics as part of the Information Technology in Organisations program. This start is far short of the level of funding required to develop this field within CS.

The North American CS curricula must also include opportunities for students to learn the most reliable knowledge about the social dimensions of systems development and use (Denning, 1992). These opportunities, formed as courses, can provide varied levels of sophistication. The most elementary courses introduce students to some of the key topics in Organisational Informatics and the limitations of Systems Rationalism as an organising frame (for example, Dunlop and Kling, 1991a). More advanced courses focus on specific topics, such as those I have listed above. They teach about substantive problems and theoretical approaches for analysing them. While many of these approaches are anchored in the sociological theory of organisations, CS students usually won't grasp the importance of the theories without numerous computing examples to work with [4]. They also have trouble grasping the character of computing in organisations without guided opportunities for observing and analysing computerisation in practice. Consequently, some courses should offer opportunities for studying issues of computerisation in actual organisations.

Fortunately, a few CS departments offer some courses in Organisational Informatics. In addition, some CS faculty who research and teach about human behaviour in areas like Human- Computer Interaction and Software Engineering can help expand the range of research and instruction. Curricula would vary, but they should include diverse courses for students who seek basic exposure to Organisational Informatics and those seek more thorough instruction. Unfortunately, only a fraction of the CS departments in the US. have faculty who study and teach about computing and human behaviour.

While the study of Organisational Informatics builds upon both the traditional technological foundations of CS and the social sciences, the social sciences at most universities will not develop it as an effective foundational topic for CS. On specific campuses, CS faculty may be able to develop good instructional programs along with colleagues in social sciences or Schools of Management.

But delegating this inquiry to some other discipline does not provide a national scale solution for CS. Other disciplines will not do our important work for us. Mathematics departments may be willing to teach graph theory for CS students, but the analysis of algorithms would be a much weaker field if it could only be carried out within Mathematics Departments. For similar reasons, it is time for academic Computer Science to embrace Organisational Informatics as a key area of research and instruction.

Notes

1. See Dunlop and Kling, 1991a for an accessible introduction to debates. Economic statistics about national level productivity are inexact, and sometimes weak. Baily and Gordon (1988) examined the extent to which measurement problems account for the difficulties of seeing the positive computerisation show up in the US national productivity statistics. They concluded that measurements were inexact, and very poor in some sectors like banking, measurement errors were not the primary cause of difficulties.
2. Organisational Informatics is a new term, and I have found that some people instantly like it while others are put off. I've experimented with alternative labels, like Organisational Computing, which has also resulted in strong and mixed reactions. Computing is a more common term than Informatics, but it's too narrow for some researchers. Informatics also can connote "information," which is an important part of this field. Sociological Computer Science would have the virtues of being a parallel construction of Mathematical Computer

Science, but doesn't connote information either. I have not yet found a short distinctive label which characterises the field and whose connotations are rapidly grasped by both outsiders and insiders.

3. Jacky's early study was based on published reports, while Leveson and Turner's more thorough study was based upon a significant body of original documents and interviews with some participants.

4. One hears similar concerns about teaching mathematics to CS students. CS students are much more motivated to learn graph theory, for example, when they learn those aspects which best illuminate issues of computation and when their teaching includes some good computing examples.

References

Baily, Martin Neal and Robert J. Gordon. 1988. "The Productivity Slowdown, Measurement Issues, and the Explosion of Computer Power." *Brookings Papers on Economic Activity* 2:347-431.

Bentley, Richard, Tom Rodden, Peter Sawyer, Ian Sommerville, John Hughes, David Randall and Dan Shapiro. 1992. "Ethnographically Informed Systems Design for Air Traffic Control." *Proc. Conference on Computer-Supported Co-operative Work, Jon Turner and Robert Kraut* (ed.) New York, ACM Press.

Boland, Richard and Rudy Hirschhiem (ed). 1987. *Critical Issues in Information Systems*, New York: John-Wiley.

Bullen, Christine and John Bennett. 1991. Groupware in Practice: An Interpretation of Work Experience" in Dunlop and Kling 1991b.

Bubenko, Janis. 1992. "On the Evolution of Information Systems Modeling: A Scandinavian Perspective." in Lyytinen and Puuronen, 1992.

Cotterman, William and James Senn (eds). 1992. *Challenges and Strategies for Research in Systems Development.* New York: John Wiley.

Curtis, Bill, Herb Krasner and Niel Iscoe. 1988. "A Field Study of the Software Design Process for Large Systems," *Communications. of the ACM.* 31(11):1268-1287.

Denning, Peter. 1991. "Computing, Applications, and Computational Science." *Communications of the ACM.* (October) 34(10):129-131.

Denning, Peter. 1992. "Educating a New Engineer" *Communications of the ACM.* (December) 35(12):83-97.

Dunlop, Charles and Rob Kling, 1991a. "Introduction to the Economic and Organisational Dimensions of Computerisation." in Dunlop and Kling, 1991b.

Dunlop, Charles and Rob Kling (ed). 1991b. *Computerisation and Controversy: Value Conflicts and Social Choices.* Boston: Academic Press.

Ehn, Pelle. 1991. "The Art and Science of Designing Computer Artifacts." in Dunlop and Kling, 1991.

Fish, Robert S., Robert E. Kraut, Robert W. Root, and Ronald E. Rice. "Video as a Technology for Informed Communication." *Communications of the ACM,* 36(1)(January 1993):48-61.

Galegher, Jolene, Robert Kraut, and Carmen Egido (ed.) 1990. *Intellectual Teamwork: Social and Intellectual Foundations of Co-operative Work.* Hillsdale, NJ: Lawrence Erlbaum.

Greif, Irene. ed. 1988. *Computer Supported Co-operative Work: A Book of Readings.* San Mateo, Ca: Morgan Kaufman.

Grudin, Jonathan. 1989. "Why Groupware Applications Fail: Problems in Design and Evaluation." *Office: Technology and People.* 4(3): 245-264.

Hartmanis, Juris and Herbert Lin (eds). 1992. Computing the Future: A Broader Agenda for Computer Science and Engineering. Washington, DC. National Academy Press. [Briefly summarised in *Communications of the ACM*, 35(11) November 1992].

Hewitt, Carl. 1986. "Offices are Open Systems" *ACM Transactions on Office Information Systems.* 4(3)(July): 271-287.

Hughes, John A., David Randall, and Dan Shapiro. 1992. "Faltering from Ethnography to Design." *Proc. Conference on Computer Supported Co-operative Work*, Jon Turner and Robert Kraut (ed.) New York, ACM Press.

Iivari, J. 1991. "A Paradigmatic Analysis of Contemporary Schools of IS Development." *European J. Information Systems* 1(4)(Dec): 249-272.

Jacky, Jonathan. 1991. "Safety-Critical Computing: Hazards, Practices, Standards, and Regulation" in Dunlop and Kling 1991b.

Jarvinen, Pertti. 1992. "On Research into the Individual and Computing Systems," in Lyytinen and Puuronen, 1992.

King, John L. and Kenneth L. Kraemer. 1981. "Cost as a Social Impact of Telecommunications and Other Information Technologies." In Mitchell Moss (ed.) *Telecommunications and Productivity*, New York: Addison-Wesley.

Kling, Rob. 1987. "Defining the Boundaries of Computing Across Complex Organisations." *Critical Issues in Information Systems.* edited by Richard Boland and Rudy Hirschheim. pp:307-362. London: John Wiley.

Kling, Rob. 1992. "Behind the Terminal: The Critical Role of Computing Infrastructure In Effective Information Systems' Development and Use". Chapter 10 in *Challenges and Strategies for Research in Systems Development.* edited by William Cotterman and James Senn. pp. 365-413. New York: John Wiley.

Kling, Rob. 1993."Computing for Our Future in a Social World" *Communications of the ACM*, 36(2)(February): 15-17.

Kling, Rob and Charles Dunlop. 1993. "Controversies About Computerisation and the Character of White Collar Worklife." *The Information Society*. 9(1) (Jan-Feb):1-29.

Kling, Rob and Lisa Covi. 1993. Review of Connections by Lee Sproull and Sara Kiesler. *The Information Society*, 9(2) (Mar-June).

Kling, Rob, Isaac Scherson, and Jonathan Allen. 1992. "Massively Parallel Computing and Information Capitalism" in *A New Era of Computing*. W. Daniel Hillis and James Bailey (ed.), pp:191-241. Cambridge, Ma: The MIT Press.

Kling, Rob and Walt Scacchi. 1982. "The Web of Computing: Computing Technology as Social Organisation", *Advances in Computers*. Vol. 21, Academic Press: New York.

Kraemer, Kenneth L., Dickhoven, Siegfried, Fallows-Tierney, Susan, and King, John L. 1985. *Datawars: The Politics of Modeling in Federal Policymaking.* New York: Columbia University Press.

Kyng, Morton and Joan Greenbaum. 1991. *Design at Work: Co-operative Work of Computer Systems.* Hillsdale, NJ.: Lawrence Erlbaum.

Ladner, Sharyn and Hope Tillman. 1992. "How Special Librarians Really Use the Internet: Summary of Findings and Implications for the Library of the Future", *Canadian Library Journal*, 49(3), 211-216.

Leveson, Nancy G. 1992. "High Pressure Steam Engines and Computer Software." *Proc. International Conference on Software Engineering, Melbourne*, Australia. (May).

Leveson, Nancy G. and Clark S. Turner. 1993. "An Investigation of the Therac-25 Accidents." Computer July. (Published in 1992 as *Technical Report #92-108*. Department of Information and Computer Science, University of California, Irvine).

Lewis, Philip M. 1989. "Information Systems as an Engineering Discipline." *Communications of the ACM* 32(9)(Sept):1045-1047.

Lucas, Henry C. 1981. *Implementation : the Key to Successful Information Systems.* New York: Columbia University Press.

Lyytinen, Kalle and Seppo Puuronen (ed.) 1992. Computing in the Past, Present and Future: Issues and approaches in honor of the 25th anniversary of the Department of Computer Science and Information Systems. Jyvaskyla Finland, Dept. of CS and IS, University of Jyvaskyla.

Orlikowski, Wanda. 1992. *"Learning from Notes: Organisational Issues in Groupware Implementation."* *Proc. Conference on Computer-Supported Co-operative Work*, Jon Turner and Robert Kraut (ed.) New York, ACM Press.

Poltrock, S.E. and Grudin, J., in press. Organisational Obstacles to Interface Design and Development: Two Participant Observer Studies. *ACM Transactions on Computer and Human Interaction.*

Sarmanto, Auvo. 1992. "Can Research and Education in the Field of Information Sciences Foresee the Future of Development?" in Lyytinen and Puuronen, 1992.

Sproull, Lee and Sara Kiesler. 1991. *Connections: New Ways of Working in the Networked Organisation.* Cambridge, Mass.: MIT Press.

Suchman, Lucy. 1983. "Office Procedures as Practical Action: Models of Work and System Design." *ACM Transactions on Office Information Systems.* 1(4)(October):320-328.

Winograd, Terry and Fernando Flores. 1986. *Understanding Computers and Cognition.* Norwood, NJ: Ablex Publishing.

Acknowledgements

This paper builds on ideas which I've developed over the last decade. But they have been deepened by some recent events, such as the CTF report. They were also sharpened through a lecture and follow on discussion with colleagues at the University of Toronto, including Ron Baeker, Andy

Clement, Kelley Gottlieb, and Marilyn Mantei. Rick Weingarten suggested that I write a brief position paper reflecting those ideas. At key points, Peter Denning and Peter Neumann provided helpful encouragement and sage advice. I also appreciate the efforts of numerous other friends and colleagues to help strengthen this paper through their comments and critical assistance. The paper is immeasurably stronger because of the prompt questions and suggestions that I received in response to an evolving manuscript from the following people: Mark Ackerman, Jonathan P. Allen, Bob Anderson, Lisa Covi, Brad Cox, Gordon Davis, Phillip Fites, Simson Garfinkel, Les Gasser, Sy Goodman, Beki Grinter, Jonathan Grudin, Pertti Jarvinen, John King, Heinz Klein, Trond Knudsen, Kenneth Kraemer, Sharyn Ladner, Nancy Leveson, Lars Matthiesen, Colin Potts, Paul Resnick, Larry Rosenberg, Tim Standish, John Tillquist, Carson Woo and Bill Wulf.

5 The Social Context of Software Design

HAROLD SALZMAN
UNIVERSITY OF MASSACHUSETTS,
USA

Introduction

Understanding computer technology as it is designed for and used in the workplace requires analysis of organisations, the labour process, and technology in general. While some outstanding work has been done on how technology and organisations interact during implementation, there is still a need for further development of a dynamic model of the social process of technology creation and use. This chapter outlines some of the important elements for such a model developed from empirical analysis of the development and use of mission-critical software systems. (An extended account of this research and analysis appears in Software By Design: Shaping Technology and the Workplace, Salzman and Rosenthal, 1994).

Analysis of information technology is too often based on decontextualising the technology, and those who use it, from the environment of use. This shortcoming is particularly significant for analysis of software that is a central or mission-critical technology in a business firm. The technology is often cast as having inherent and immutable characteristics that determine the impact it will have when used in the workplace. "Computer impacts" are thought to be independent of the organisational (and broader social) context in which the computers are used. The decontextualisation of computerisation can also be seen in the term "user," which reduces all people in an organisation to one common, and not particularly salient, denominator (e.g., they are also as likely to be "drivers" of automobiles and "users" of pens as well).

The reference to "user" levels all the important distinctions of function and position of people as they do their work.[1] The use of a computer is often regarded as reducing all workers to a common status absent of any social or organisational context that defines them. Recognising differences in power, position, function, and organisational structure are all central in analyses of the workplace or society. These are dimensions of the social context that are important for analysis of computerisation in general and in the design of software in particular. This chapter examines the social context of software design first in a general, theoretical perspective on technology and then examines the specific context of software design in the workplace.

Part I: Perspectives of technology

Analysis of the social context of software design and use is the basis for a perspective of technology as socially shaped. The premise that technology design is socially shaped may be acknowledged in very general terms but it is not taken to be an important consideration in the "real" tasks of design by those within the design community. Moreover, as David Noble (1984, p. xii) has observed, "although it has belatedly become fashionable among social analysts to acknowledge that technology is socially determined, there is very little concrete historical analysis that describes precisely how". Although Noble's observation is still largely true, since his pioneering work there has developed a growing interest in and body of research on the social shaping of workplace technology.[2]   However, few studies examine the role of engineers in creating technology and specific technological designs.

The predominant view of engineering is that it is "applied science", that is, it is the application of scientifically and objectively determined principles.   The "scientific view" of technology is that advances in knowledge are largely independent of subjective influences. Thus, technology reflects engineers' calculations of the most economic and efficient designs to utilise that knowledge. This is the dominant view of engineering as expressed, for example, in the U.S. by the accreditation Board for Engineering and Technology definition of engineering as "the profession in which a knowledge of the mathematical and natural sciences gained by study, experience, and practice is applied with judgement to develop ways to utilise, economically, the materials and forces of nature for the benefit of mankind" (quoted in Thuesen and Fabrycky, 1989). Insofar as social choices or values are considered, they are regarded as important for decisions about the use or development of a technology but not as an integral part of the design process.

In opposition to this position is the critical social perspective in which technology design is cast as the reflection of a society's political-economic structure. Social science analyses of technology traditionally have been from a political-economic perspective. Political-economic perspectives of technology and organisation are largely focused on workplace technologies with the technology viewed as one of the major, if not the sole determinant of social organisation. In the nineteenth century, political economists such as Marx and Smith viewed technology as a driving force shaping the political-economic system with little attention paid to how social forces might be shaping technology.

More recently, social scientists and historians have examined the "social" character of technology. David Noble (1984, 1977), who contributed some of the first analyses of the social forces shaping engineering designs, challenges the idea of an inner logic driving technology development and leading to specific designs. Tracing the rise of engineering as a profession within the confines of industry (in contrast to other professions which developed as independent practices), he finds that, as a result, the values of engineers reflect those of their employers, only marginally distinct if at all, from managerial objectives. Engineering work, Noble concludes, is oriented toward developing technology that reinforces the existing political and social order.

In recent years a new perspective has been developing based on a more dynamic view of the social shaping of technology. The guiding premise in this work is that technology is socially shaped or constructed and part of a larger network of things and people. Using this framework, sometimes referred to as a social construction of technology perspective, and building on the traditional studies of science, technology, and society, a number of studies have examined the ways that technology decisions are shaped by nontechnical factors. Research within the emerging field of the social shaping of technology varies quite dramatically in the approaches used, especially in defining the relevant range of social factors considered relevant or salient.

At the microlevel, some researchers focus on how people who use the technology interpret or define the important characteristics of an object. In the extreme, some researchers view technology as having no inherent properties but as essentially open to interpretation by individual actors (e.g., it is in the eye of the beholder whether a personal computer is just a faster typewriter or a means for document composition in which typing is incidental to its use and in which the technical features of the computer are virtually irrelevant). Implied in this view is that technology will not constrain users in accordance with an embodied set of values because technology impact depends on how the users interpret it.[3]

Although some research in this area tends toward a view that technology impacts are entirely the result of users' perceptions (and thus technology, per se, is not a relevant subject of study), these researchers do make an important point that we need to examine technology not just as an assemblage of artefacts but also in terms of the meanings people find in its use. A common thread uniting the spectrum of perspectives on the social construction of technology is the challenge to the technical view in engineering and the hard technological determinism view in the social sciences and humanities.   Building on this perspective, we view the engineer and technology designer as actively shaping technology and reflecting the values and interests of their social environment.

In developing a theory of software design from initial design through redesign in implementation, we need to consider several different dimensions of the "social" process of technology design. First, what is the role of designers as individual creators of technology? Second, what are the important factors of the organisational context of technology design and use? And, third, how do we assess the role of the "user" in specifying software requirements?

The Engineer as technology designer

The engineer's role in shaping technology is seen as either an objective agent applying the laws of science or as a subjective agent shaping technology in accordance with individual whims and values. A more nuanced view regards the engineer as working within an organisation that directly and indirectly imposes constraints on his or her work. Thus, values and subjective decisions that shape technology are mediated but not wholly originated by the engineer. It is this latter perspective we find useful for understanding the social shaping of technology and the limits of influence by individual designers.

The engineer, in his or her role as creator, works with external social as well as technical constraints. Just as every artist must play to an audience, so must every engineer. Although the engineer, like the artist, is the actor conceiving and creating the artefact, it is not an immaculate conception. Individuals both mediate and reflect values shaped by organisations and by the society at large. Even when new forms of art develop, their creators are often viewed as giving expression to new ways of thinking or new developments in the world. They may be considered pioneers of a new form but it is usually related to changes in a broader context and rarely considered sui generis. Artists must often play to the art critic for success because the critics often wield the power to decide the fate of each artist's masterpiece. (If they go beyond the frontier of accepted expression they may find success too late for them to enjoy it.)

At the same time we know that the creation of artefacts is not just a response to the status quo but also reshapes existing structures and values. Again by way of analogy, language has similar properties of shaping and being shaped by the action of individuals.[4]   We are educated and socialised to structure our expression according to very specific and highly prescribed rules of grammar. If this were a unidirectional process, language would never change. Language is considered "living" in the sense that it does change in response to changes in forms of expression by individuals who are, at the same time, expressing themselves largely in accordance with existing rules of grammar. The change does not usually reflect intentional decisions by individuals but the congealing of a number of factors that both give birth to a new form and sustain it to a point of acceptance (e.g. the need for a new word for a new technology or a new way of thinking about the world).

The design of technology is a process with similarities to the development of language and to the development of new art forms. It will reflect a prevailing set of values. Individual designers may give expression to these values but they have limited latitude to introduce into the technology new values or values that are incompatible with their environment. Engineers must generally play to the preferences of those wielding power. Unlike artists who may achieve posthumous success for creations unappreciated in their lifetime, the creations of engineers may not see the light of day if they do not receive approval. Engineers in the modern world work in organisations that are subject to the vagaries of the marketplace. An engineer's superiors have the power to quash errant designs or the marketplace will provide the discipline, sometimes by eliminating organisations that are unable to develop effective internal controls.

At the same time, engineers do mediate the preferences of others when they interpret and translate them into a material technology. To some degree, designers can push the boundaries of existing structures by designing technology that becomes one factor reshaping those structures. Although the enlightened designer can push the boundaries a little faster and a little wider, significant change is not one of individual action but the consequence of a broader set of factors that must congeal to shape and support new design approaches. In summary, engineers neither unreflectively implement directives from above and outside nor do they spontaneously and independently shape technology.

We need to consider the environment in which engineers work and the environment in which technology is used. The environment or boundaries shaping and regulating the activity of designers and users may be defined as the "design space" of designers and the "action space" of users. Conceptually this notes two domains of action that are the amalgam of a number of different pressures. These are heuristic concepts, not entities with clearly demarcated boundaries (in contrast to, for example, a company which is an entity that is distinct from its environment). These are the "virtual spaces" in which organisational objectives, market pressures, professional training, technology, and the entire host of factors in the social and material world form the immediate environment individuals and groups experience Thus, the impact of a technology will be experienced as it interacts with other elements in the user's immediate environment. To some degree the action space might be seen as constructed by the user (e.g. to the extent that a person interprets organisational rules as a consideration in his or her actions), but it also has certain known, consistent, and externally imposed properties (e.g. certain actions rather predictably lead to discipline from supervisors or malfunctions in a machine and the power of some can be exercised to influence the action of others and the outcome of decisions).

Focusing specifically on workplace technology provides a specific context for analysis of how the technology is socially shaped to mediate and regulate the work process. Features and functions of a specific technology or system are designed to operate in ways that shape the work process in a particular way, in accordance with a particular set of values. The technology is not deterministic in the sense that actual use can be deduced from design. However, the properties of the technology do provide constraints on, and impediments to, the scope of action of users; alternatively, their design can increase the scope of actions by users. At the same time, the organisation of the workplace is a factor shaping the values guiding technology design.[5]

Thus, analysis of software design requires an understanding of the organisations in which designers work and organisations in which the technology is used.

The organisational context of design

To say that individual designers work in an organisational context requires that we examine design at the organisational level. The extensive bodies of literature in the sociology of organisations suggest that there are common and enduring characteristics of organisations that exert systematic influence on the activities of people in them. Analysis of these relations and social shaping of software is informed by the insights in structuration theory as developed by Anthony Giddens and others.[6]

Just as technology embodies values in its design, decisions in organisations are influenced by values and are not made solely according to a neutral, rational logic to achieve a commonly shared set of goals. Organisations operate as cauldrons of vested interests, calcified patterns of operation, and in many ways as "contested terrain" over which different groups vie for control. Central to organisational dynamics is how power is used in the organisation. Decisions do not necessarily represent an optimal choice among alternatives or even one that is most effective for the organisation or those it serves. The marketplace does serve to constrain the irrationality and ineffectiveness of organisations to a degree, but it is a very loosely coupled system as attested to by the survival of many inefficient organisations for long periods of time.[7]

Drawing on sociological perspectives of organisations, there are several social factors important in understanding the dynamics of technology design and use. Clearly, organisational decision making is constrained by "bounded rationality". In the search for the optimal solution, not all information about possible options and consequences is available to those making the decisions. But, the constraints are not just technical but also political. Organisations have multiple and conflicting goals. Even the overarching goal of an organisation generally does not bring together all of its constituents and their interests. Different people and departments also have competing and conflicting goals. Understanding power and authority in the organisation is central for analysis of how these conflicts are mediated and how preferences and objectives are reflected in decisions.

Power in organisations

Decisions are influenced by the power different functional groups have vis-à-vis other functional groups and by the power vested in hierarchical differences of the groups and people involved. That is, power is held at a horizontal level, often by coalitions, and also in vertical, hierarchical form that tends to be stable and have characteristics common to many organisations. Power can also be differentiated as control over resources and as control over people in the organisation. This is the difference between economic and political power and authority in the organisation.

At the horizontal level we can examine power holders in an organisation by dominant coalitions or groups, or particular to certain types of decisions, such as a marketing department having greater decision making power over product design than an engineering group, or a computer department having more power than users in decision making about computer equipment and software acquisition. This type of power can thus change with realignment of coalitions, it can depend upon the type of decision at issue or from changes in control over resources, and may vary from organisation to organisation.

From another perspective power is an attribute of hierarchical organisation so that level in the organisation indicates clear subordinate-superior relationships. This may also be considered a function of authority, that the superior relationship commands more expertise and is assigned greater responsibility over broader realms of decisions. However, our focus is on the uses of power in these arrangements, over the control of resources and the influence that is based on this control of resources and that leads to some preferences prevailing over others. In traditional work organisations there is generally a chain of command and a generally consistent realm of power and authority at each level. The power in the hierarchy tends to be relatively stable over time and in here in the basic structure of the organisation, whereas horizontal types of power may be products of specific coalitions or control of resources. Decisions about work process, for example, are traditionally regulated in a hierarchical manner. In recent years there has been a fair amount of attention to the need to change traditional hierarchies and decision making, but this has resulted in changing only a small minority of work organisations in the U.S.[8]

Organisations and innovation

There is a long-established literature on how technology shapes organisations, but in recent years researchers have significantly qualified the degree of determinism in this relationship and some have severely criticised earlier research on technology and organisational structure.

Although a simplistic technological determinism is generally not an explicit foundation of current theory on technological innovation and organisational form, some researchers find it continues as an implicit factor. As Child, Ganter, and Kieser (1987, p. 99) note, "The failure to address this process [the interaction of technology and organisations] in effect abstracts technology and organisation from their contexts and in so doing encourages the naive expectation of a mechanistic relationship that has infused a great deal of organisational research". Scott's (1988) review of the literature on technology and structure concludes that the most promising theoretical perspectives are those that recognise the "dynamic properties of organisational structure". Scott suggests that the most fruitful technology-organisation studies will be those viewing "structure as process" (building on Giddens' structuration theory, for example). Scott (1988, p. 29) points to research studies in this area that "emphasise the role played by organisational politics, by vested interests, by institutional arrangements in shaping and selecting technologies and in designing structures". Scott and others are critical of traditional technology and organisation studies for often examining only one part of the technology-structure dynamic.

The dynamic of organisational structure and technology adoption, use, and design and the attributes of organisations that shape this process is the important part of this body of research for examining software. Because organisations tend toward stasis as a way of perpetuating themselves and as a way for power holders within the organisation to maintain their positions, one tendency is for organisations to resist technology that may lead to restructuring or disrupt the larger network and environment of which they are a part. As Hughes (1987, p. 57) observes, "Because radical inventions do not contribute to the growth of existing technological systems, which are presided over by, systematically linked to, and financially supported by larger entities, organisations rarely nurture a radical invention". Evidence from a study by the economist W. Paul Strassman (1959) lends support to this point in finding slow rates of technology diffusion. He found that for 10 to 15 years after a new technology came into use, the old obsolete technology continued to not only be used but to grow (e.g. in power generation and steel production). He concludes that many firms will not abandon current methods and technologies so long as old technologies and production methods are just profitable enough, even if not as profitable as new ones. A related phenomenon has been examined by others (e.g., David, 1985) as technological "lock-in". New technology may not replace existing ones that are locked-in through adoption and use by many organisations even if the new ones are much more profitable.[9]

Organisational factors impeding technology innovation and diffusion have been discussed by Child, et. al. (1987, p. 99) as organisational conservatism.[10] Understanding the innovation process, they argue, requires "a theory of the relation between technology and organisation. This theory, however, remains underdeveloped, particularly with respect to the process by which technological investment, design of systems and organisation, and implementation are decided, and the factors impinging on this".

These analyses of acceptance and rejection of technology in organisations address a dichotomous adopt-or-reject decision about a technology based on its properties as designed. They also generally focus on large-scale technology changes such as steam versus electric power. However, the dynamics of organisational conservatism also affect the process of technology adoption and its redesign during implementation and use. Adoption of a radical technology in an organisation can be viewed as occurring through a process of assimilation and accommodation. Borrowing from Piaget's (1967) concepts of how children learn and his overview of structuralism (1968), we posit an analogous process of organisations assimilating or accommodating new technologies (this occurs at both the individual user level and at the organisational level as predominant practice).

A new technology may be assimilated into existing structures and work procedures so that it is used in a manner equivalent to previous technologies. In this way existing frames of reference may be imposed onto the technology and the impact of the technology will be minimal. Users and organisations will continue with only slight modification of previous patterns of technology use and organisational operations. Accommodating a new technology is a process of restructuring the organisation with the adoption of new technology, sometimes to change the organisation and work processes in ways that utilise new capabilities of the technology. Sometimes the technology provides the impetus for organisational restructuring that is related to many other factors as well.[11] However, there is a tendency toward assimilating technology rather than accommodating it.

In both cases the organisation shapes the technology as it incorporates it into its structure. A personal computer, for example, may be assimilated into the organisation and used as just a faster typewriter or there may be accommodation in which the activity of typing changes into document composition.[12] Assimilation and accommodation describe the process of organisations adopting technology as dynamic rather than deterministic.

In summary

Technology design, adoption, and use are interrelated in an ongoing process mediated by structure. "Structures" of both tangible technologies and intangible organisational norms and procedures shape action and are, in turn, shaped by actors. Moreover, the processes of design and use are embedded in organisations which, in turn, are embedded in the structures of their markets, which are shaped by broader characteristics of the society. It is a loosely coupled system, so that the structures in each context exert influence on design and use of technology but these are also mediated by the actors and structures at each level. While technology embodies organisation, organisations and their environments also shape technology. The organisation/technology division artificially separates this ongoing and recursive process. "Technology" thus represents a particular moment in history during which the technology as artefact takes form. The same technology is reinterpreted and redesigned at later stages reflecting changes in its environment of use. This occurs as a seamless dynamic process, with less separation of stages in the life cycle than appears in a formal retrospective analysis.

Thus, analysis of technology should focus on how people design and use the technology while also realising that they both mediate and interpret a series of social influences. That is, decisions and actions taken by individuals are shaped by the social context in which they operate but individuals also push the boundaries of existing structure in ways that reshape those structures.

With this general theoretical background, it is important to examine the specific context of technology design and use. In analysis of software, particularly mission-critical software used in the workplace, it is important to see the "user" in his or her role as worker.

The "User" as "Worker"

Unlike the home user of a personal computer, who is working toward self-defined goals, the user of large systems in the workplace is using the software in his or her role as "worker." Thus, generic discussions of users must, at least in these applications, be related to what we know about workers and work. It also should be noted that most people considered "users" by software designers would rarely identify themselves that way. Rather, they would define themselves in terms of their job as, for example, a bank teller, a manager, or an accountant.[13] This is particularly important because the software user literature fails to address issues that are widely discussed elsewhere about workers in organisations.

Particularly important in workplace software is that, because it is designed for multiple users, across department lines and hierarchical levels, the system impacts are uneven. Computer systems generally mediate work activity to a greater degree for lower level workers than for managers and professionals. At higher levels, users depend on the software for crucial information but their actual work routines are not regulated by the computer. A bank manager or airline company accountant will use computer-generated information, but it is the bank teller or airline reservations clerk whose essential and ongoing activities depend on the system. A system crash may interrupt some work tasks of a manager or accountant, but there is a large portion of his or her work that can still be performed; for the bank teller or airline reservation clerk a system crash leaves them idle. Similarly, a poorly designed feature that is an inconvenience for the home PC user or the manager may be a significant impediment for effective work to the clerk. If it is a feature or function designed to control the flow of work or to regulate the worker, it, too, will be more acutely felt by those at the bottom.

Not only is the perception and impact of technology different at different levels of the organisation, but interests and goals of workers will vary at different levels of the organisation. One premise of management theory, from Taylor to contemporary human relations theorists, is that diverging interests and conflicts are not inherent to work organisations, they just arise from bad management or poorly socialised workers. Conflict is thus seen as a problem of managing a recalcitrant workforce not as the need to negotiate legitimate and rational differences. In a classic textbook on industrial sociology, Miller and Form (1980, pp. 124-126) compare the organisation "to a machine that requires the individual to be mated to it," in which "every person in it must be moulded to some degree into the image of the organisation," in which there is a "fusion" between the individual and the organisation which "tends toward equilibrium." This is considered the normal and desirable state of individuals in an organisation.[14]

If conflict or lack of fusion between workers and organisational goals are seen as idiosyncratic, then they become peripheral to consideration of how a system should be designed. If the differences are viewed as persistent but irrational, or rational but illegitimate (e.g., "workers who just do not want to put in a 'fair day's work for a fair day's pay'"), then an objective of technology design may be seen as a means of increasing managerial control to maintain effective operations. This objective in software design is then viewed as not only legitimate but also as not being problematic. That is, software can be designed to help managers better accomplish what needs to be done and improve the rationality of operations without engendering significant resistance or inefficiencies.

This model of the workplace is not only inaccurate but does not lead to effective or desirable methods of software design. Rather than organisations operating on the basis of consensus around a common set of goals, organisations operate to mediate diverging interests. This is not a view universally accepted in management theory or by technology designers.

Another important factor for software design is the divergence between the formal description and the reality of workplace operations (e.g., Weltz, 1991). Formal procedures, policies, and rules are not likely to represent the way the organisation actually functions or how tasks are actually done. In fact, some researchers argue that organisations may need a loose coupling between the formal rules regulating procedures and actual practices.[15] This reflects the tension between the generation of bureaucratic rules to regulate procedures and the realities of the world that require flexible action. Informal work routines are not formalised in part because of their very nature (i.e., being ad hoc or at least not officially sanctioned) and in part because formal acknowledgement of nonstandard procedures might threaten the legitimacy of hierarchical and bureaucratic organisation.

Insofar as software mediates work tasks and procedures, the designer needs to confront the longstanding issues of workplace conflict and regulation as part of the design task. Perspectives of work, therefore, play an important background role shaping the definition of requirements for design and, in the user organisation, shaping the action space of workers.

Studies of technology and regulation of the workplace suggest that historically one requirement has been to provide means of control to support managerial attempts to better regulate the work process.[16] To the extent that this occurs, the technology designer plays a role beyond the automation and regulation of tasks, moving, perhaps unwittingly, to the regulation of people. This perspective of work identifies issues far different than the typical consideration of the requirements of the line "user" in software design. Effective design must consider the role of work in people's lives and the role of the user as worker.

Part II: The social shaping of software

Design for the user: A limited content

User-centred criteria for the design of applications software, unlike those for hardware design, have been explicitly promoted in design methodologies.[17] In fact, it was even declared that, "The 1980s could go down as the decade of the user."[18] Design for the user was generally described in terms of user-friendly interfaces or, more generally, of providing features and functionality that make the system effective for the user. Nevertheless, systems continued to be designed that were ineffectual for users and/or that users would not use. A widely prescribed remedy became user involvement in development, a "user knows best" solution. Some argue that only by involving users in design can the designer ensure that the software is effective.

This nostrum for curing system deficiencies has widespread currency in both the research literature and textbooks and has a great deal of intuitive appeal. It is also very much in the spirit of current thinking about the customer's role in effective design of tangible products in general. However, with respect to software, the evidence that this approach is sufficient to improve end-user effectiveness is not compelling.[19] Examining the assumptions underlying user involvement and "better design" formulae provides some insight into their inherent limitations.

The U.S. perspective on the role of the user in software design is perhaps the most limited when compared to the theoretical perspectives developed in Europe. In a review (Salzman and Rosenthal, 1994) of 50 textbooks on applications software design, we examined the nature of user considerations in design methodologies.[20] We first looked in the index for listings of "users." Twenty-eight of the 50 books had at least one such listing. We analysed the sections in each of these books that referred to users for the types of user consideration in design that were being promoted.

(These textbooks were all written by U.S. authors. Some of the European researchers have approached software design quite differently, as discussed below.) Of the 28 books that had an index listing of user, only 17 had more than a passing reference to user consideration in design. These 17 books all note the importance of designing software that meets user requirements. Only six books mention specific reasons for difficulties in designing user-effective software. In our review, these and the other textbooks propose four types of solutions: talking to all users, having an analyst or user liaison whose specific job responsibilities include determining user requirements (usually by formal user interviews), having frequent dialogue between designers and programmers and users, or even putting a user in charge of software development projects.

Several limitations in these various solutions relate to the background assumptions in the design approach. One such assumption is that ineffective design is just a technical problem of inadequate methods for specifying or assessing user requirements. The unstated belief here is that there are optimal solutions to user needs and that efficiency and effectiveness can be objectively evaluated with the proper methods. Resulting failures or shortcomings in designing an effective system are, therefore, thought to stem either from poorly defined specifications (i.e., as a methodological issue involving better surveys, more representative and active participation by users, etc.), or technical limitations (e.g., tradeoffs or constraints imposed by hardware or software technology). In theory, there are no inherent limits to achieving an optimal design solution that meets user requirements.

The conclusion of these texts, sometimes explicit and other times implicit, is that designers also do not need to be concerned with learning about or considering the social or organisational context in which the software will be used. There is no framework provided for systems designers to understand social and organisational implications of their work, such as that provided in some of the organisational literature. Rather, the prescription of involving users in the design process is assumed to be sufficient to ensure that systems are designed to be effective.

In general, the limitation of these standard approaches is an overly technical and narrow focus on the context of software use and role of the user. It divorces technology design from use and focuses on the creation of software as just an abstract enterprise.

These U.S. studies tend to address discrete dimensions of organisations, power, and the user.[21]   For example, power in organisations has been addressed in a number of studies that examine the use of computer systems to strengthen the power of certain groups.[22] Studies addressing organisational aspects of software design also concentrate on the fit between the user organisation and the specifications of the computerised system. The central focus of these studies is how computer systems may alter power relationships and control of information in the organisation.[23] Some of the problems of power are identified as issues of expertise, of the monopoly of power through professionalism and technical mystique. For example, domination of technology design and use may be viewed as computer professionals exercising their body of knowledge and technical expertise to gain organisational power and fulfil their particular needs such as empire building. Thus, there is a focus on democratising knowledge through user participation in software design as the countervailing force to power in the organisation. The limitation of increasing user knowledge without increasing their power in the organisation is generally not addressed.

This U.S. approach deals with specific and discrete aspects of organisational life and technology. This may be seen as consistent with a historical tradition of examining issues in technical and pragmatic terms, with issues of politics and power being subsidiary. Workplace theories and approaches to work reorganisation, for example, tend to focus on techniques of management, not examining systemic issues of hierarchy and power. In terms of software this general perspective shapes both analysis and the subject studied. For example, issues of power tend to be seen mostly as the monopoly of power by groups or coalitions, not the type of power that represents different "classes" in the organisation (just as in the broader society, class is not an important concept in the U.S., particularly as compared to European perspectives and politics[24]).

The U.S. perspectives, in summary, may discuss power as important, but it is seen as residing in individuals and coalitions; thus, power is viewed as situational or temporal. In part this reflects a focus on the conflicts at the horizontal level of the organisation (e.g., departments that are different functionally but hierarchically relatively equivalent, such as accounting, sales, and the Management Information Systems departments). This perspective provides an important but incomplete part of the analysis. Power and conflicts are also attributes of hierarchy and particular forms of management and thus inhere in the structure of organisations. They are less fluid than some of the analyses would suggest. Vertical relationships tend to be more stable and tend to have specific, core objectives around issues of management and work organisation.

The U.S. perspective is a very pragmatic approach in software design research and theory that reflects the reality of the workplace and markets in the country. The designer, or end-user, is not likely to have much effect in reshaping the workplace. Workers have rarely been able to exercise significant influence in structural organisational change and other types of workplace reforms. To suggest that, somehow, in the context of traditional organisational structure, a more democratic or participative approach to technology design will occur seems not to reflect the broader context of technology design and use in the U.S. Among the different studies of computing in the U.S., we can find all the elements of the broader scope of organisational and worklife issues we discuss, but it does not coalesce into a coherent body of theory about computerisation and organisations.

A different approach comes from the English researchers based on longstanding traditions in studies of work in the sociotechnical school (Trist and Murray, 1990). A key premise of sociotechnical theory is that work and technology should be designed to improve worker satisfaction, which will, in turn, lead to higher levels of productivity. Most of this work focuses on job redesign to increase the meaningfulness of the job and provide workers with broader areas of responsibility and autonomy.[25] Developing this line of analysis specifically for software, Mumford (1981) suggests that software design decisions reflect the value perspective of software designers and she evaluates the extent to which several systems incorporate human, technical, and business values (drawing on McGregor's Theory X and Theory Y typologies. Design effectiveness is evaluated as the extent to which systems are designed to restore and maintain organisational equilibrium, drawing on a Parsonian model of organisations and social relations. Developing shared values among designers, managers, and system users, and humanistic values in particular, is the suggested goal for effective system design. Design that reflects sociotechnical system principles is assumed if values can be aligned among all relevant parties. Focusing on the social side of technical systems, the technical system is relegated to backstage in sociotechnical analyses. Most of the focus is on specifying user participation and advocating more concern by designers with workplace issues (along the lines of the sociotechnical framework for work redesign). The empirical studies of sociotechnical design of computer systems, notes Olerup (1989, p. 60), "give very detailed accounts of the work design process, but next to nothing on the process of designing the computer system."

In a historical review of sociotechnical theory, Mumford (1987) notes that very little of this research addresses technology design and suggests that one reason is the lack of engineering expertise by sociotechnical researchers (see also Cherns, 1987). Moreover, as Bansler (1989, p. 12) has written, "Although many of the socio-technical ideas and recommendations have gained widespread acceptance among systems designers they have, however, only had little impact on how the systems actually developed..." Although the importance of job satisfaction is a cornerstone of sociotechnical theory, in practice sociotechnical researchers have not successfully convinced managers and designers to attend to the human factor in design. Sociotechnical theory provides principles for humane work organisation and, in that way, provides indirect criteria for technology design.

Approaching the issues of technology design and use from a very different perspective are the new social constructionists such as those in England and France. Although perspectives on organisations and jobs differ, these studies follow in the same path as sociotechnical approaches in viewing the design and use of technology as the dynamics occurring within the scope of the user or individual designer. That is, design and use issues are largely, if not wholly, contained within the design space and action space. (In this view, the technology that emerges from designers is a subjectivist creation that seems to have few enduring attributes because the technology can be largely, if not completely, reinterpreted by the user acting within his or her action space. In each realm, technology design and technology use, individual values and perceptions are the determining factors). These studies provide a number of important observations and insights but, because they tend to reduce technology-related issues to highly individual values and perspectives, do not examine the role of organisations and other structures on technology design and use.

A different perspective emerges in the Scandinavian studies, which build on a sociotechnical perspective but take it to the organisational rather than the individual level. The early studies in Scandinavia identified a number of organisational level issues affecting systems development and implementation, shifting the focus of analysis of previous studies (e.g., Bjorn- Andersen and Hedberg, 1977, p. 125). In these studies, researchers concluded that improved designs will result from better communication between designers and users about "the value system rather than the actual computerised system" (Hedberg and Mumford, 1975, p. 58) and the "political aspects of designing" (Bjorn-Andersen and Hedberg, 1977, p. 139). In terms of the research in the 1970s and early 1980s, these studies provided the groundwork for developing a broader perspective.

Recent approaches in Scandinavia have tried to move beyond isolated participation of workers (users) in design, when it was found that user participation was ineffective if not based on a broader participation in workplace decisions.[26] Thus, the design approaches being advocated are based on a more far-reaching concept of workplace democracy. A number of design projects have been undertaken in Scandinavian countries to develop new approaches to design, to overcome what they view as limitations in existing theory and practice. It is very much a practical orientation, from which theory was then developed, and it was shaped by the culture and social structure of those societies. Significantly, the strong role of labour unions in those societies and corporations, the emphasis on, and government support of, quality of work life programs and a strong craft tradition provided a different atmosphere for developing design methods and approaches.

Pelle Ehn (1989) has provided the most detailed articulation of the various dimensions of these approaches. He criticises prevailing computer systems design approaches in several areas, including the basic philosophy of engineering as bound by an overly narrow focus on design as a natural science. Rather, Ehn proposes a view of design as being craft or art as much as science and as crossing disciplinary boundaries between natural sciences, social sciences, and humanities. Perhaps what most distinguishes some of the Scandinavian approaches is that they see software design as supporting an explicit goal of industrial democracy. This perspective, then, frames their approach to the design process, namely participation by trade unions in a "collective resource" approach. The design process is structured to be compatible with the organisations of many Scandinavian companies in which there is a high degree of unionisation and various forms of industrial democracy or union participation in management issues. Ehn and others address not only the design process but also the content of design, arguing that software design should support and allow for development of craft skills.

The objectives in a design approach reflect the larger organisational and social context in which it is developed and used. The contrast of design approaches in different countries illustrates how different factors are emphasised and how they reflect differences in organisational and social structure.[27]   The differences in workplace organisation can be said to reflect differences in industrial culture. That is, the combination of a country's or firm's economic market, employment practices, education system, legal system, social and engineering values, and other factors shape the way it will organise its production process or, for our purposes, technology design.[28]   In the same way, proposals for changes in technology design processes and objectives will be constrained by the environment. The cross- national comparisons of design strategies illustrate both the possibilities and the constraints.[29] Building on this research we outline a model of technology design that includes the broader context of the organisation- technology web. Examining software design in context requires a perspective of how software functions as part of the institutional web of technology and organisations that shapes the user's "action space" and how software, as a product of an institutional web, is itself shaped by the "design space" in that web.

The organisational context of software design

There is a small but growing body of research that examines the broader organisational context of software use and views technology as one part of an interconnected web of the organisation. Some of the initial theorising about a new model of computers and organisations has been formulated by Rob Kling (1987; 1991), building on the insights from a number of different research directions. As outlined by Kling (1987, p. 3, 1991), there are two underlying models in studies of computing: discrete entity models, which focus on "relatively formal-rational conceptions of the capabilities of information technology and the social settings in which they are developed and used" and web models, which are "a form of 'resource dependence' models [that] make explicit connections between a focal technology and the social, historical, and political contexts in which it is developed and used."

The discrete entity studies, such as those previously discussed in our review, have a bias toward technological determinism. The underlying perspective assumes direct technological impacts with little consideration of the organisational structure, the web composed of the social relations, infrastructure, and history of the organisation. Consensus among users within the organisation is generally assumed in discrete entity studies and the locus of systems failure is limited to developer-user interaction.[30]

Another stream of literature, characterised as web or organisational model studies, takes a broader view of different user interests and organisational contexts.[31] These writers do not assume consensus or rationality in organisations. They emphasise analysis of technology in context and social aspects of technology development and use. These web models focus on three elements: the social relations between participants, the infrastructure available for their support, and the history of commitments made in developing and operating related computer-based technologies (Kling, 1991, p. 5). Building on this web model, we include a perspective on the environment of organisations, how they are shaped and how they shape the actions of their constituents, and links to the market and broader socioeconomic environment.

The "seamless web" of technology, encompassing the full range of factors from individual values to the structure of the society in which it is created, may be so all encompassing as to include just about every social and technical factor. For our purposes we draw the net in a bit closer and sharpen our focus on a few dimensions of this larger web (see figure 5.1). Our model begins with the design space of the designer and his or her immediate environment. The values and decisions of the designer are significant in defining the technology. Thus, understanding the design as an expression of the designer is important, of seeing how individuals and design teams make decisions about design. These explanations need to be examined in context. (Designers may or may not explain their decisions as shaped by a broader set of factors, because they may believe they are determining a strictly technical set of specifications.)

The economy, Markets, Social Trends

Organization

People Working
with
New Technology

of the Workplace

and Other External Forces

Figure 5.1   The web of technology

Expanding outward within the organisation, are relations with other functional departments and hierarchical relations. Important here are the findings of organisational research on power in organisations and the limits to rationality in organisational decision making, as just discussed. The organisation is not isolated but affected by its environment and thus it is also important to examine the external environment of the software design and user organisations. In the design organisation factors that are important include pressures for competition with other vendors and selling its product to the user firm. In the user firm there is a similar range of factors. The action space of the user is shaped by the technology and other aspects of his or her position (e.g., degree of autonomy, responsibility, nature of the work tasks, etc.). The user's action space is also shaped by the overall type and structure of the organisation and the way hierarchical and functional differences are negotiated. The user organisation's market, as perceived by both the user firm and by the software designers, will shape design decisions and the viability of both the vendor's product and of the user organisation's efficiency, effectiveness, and profitability and survival.

The politics of software design

Organisational politics are crucial in the early phases of technology development and provide opportunities for those in positions of power in the user organisation to exercise the most explicit influence. Furthermore, past technology and organisational choices form patterns that are institutionalised and form the structure shaping current technology choices (cf. Kling, 1987, 1993; Thomas, 1993).[32] Thus, the initial stages of technology definition provide partial constraints on the action of users when the technology is implemented. The late life cycle stages of design are the result of a continual process of actors interpreting and negotiating the technology design and use within structural bounds of hierarchical power, resources, authority, and autonomy. Thus, one must also examine the subsequent interpretative or subjective construction of the technology by the user, although that, alone, is not sufficient.

The politics of the design process involve the politics of the user organisation and directly affect the content of technology design. In their study of the way software systems requirements are determined, Robey and Markus (1984, p. 9) advance this political perspective of technology design. They find that "systems development activities may well serve purposes beyond the rational goals of system quality and user acceptance." In case studies (e.g., Salzman and Rosenthal, 1994), it can be seen how a software system has multiple users who have conflicting interests and differing concepts of what the system should accomplish. A consensus about the purpose of new software systems may not be possible. Because a single software design is required, there is often a bias in requirements analysis reflecting the interests of those with the greatest power. Thus, there is an inherent tendency for "new" software design to maintain existing arrangements in opposition to significant change.[33]

The political nature the software design process is often not apparent. Robey and Markus (1984, p. 12), for example, show how the design of software follows certain rituals such as the formal procedures of information requirements analysis or user groups to show rationality in the process, to provide at least the appearance of objectivity: The rituals of systems development perpetuate the prevailing ideology of rationality and provide an acceptable cover for inexpressible political motives in the dealings between users and designers. Overt conflict and manipulation are thereby controlled, lending stability and order to systems development. In effect, the rituals of systems development enable participants to act in their self-interests without discrediting the organisation's rational ideology.[34]

Those in power often have conflicting goals vis-à-vis their subordinates in the organisation. Rituals or formal reality come to bear not only in the designing of the software but also in governing the environment in which the software is procured and then used. The important point in this analysis of software design is not only that it is a political process, but that the politics of software (or technology) design tend to have certain systematic outcomes.

In analysis of the software design process, Salzman and Rosenthal (1994) present case studies showing that the objectives and interests of some groups will prevail over the objectives and interests of other groups, as reflected in design specifications. This is not a politics of pluralism in which different groups all vie for power by mobilising control of resources and authority and which will shift, for example, as organisational needs or requirements change. It is not freely contested terrain in which all comers are on equal footing.[35] Thus, software design should be viewed as a political process in which the interests of the existing power structure will dominate. This does not, however, determine the final impact of the technology because of changes that occur later in the design life cycle and because of how the technology interacts with characteristics of the organisation.

The dual reality of organisations

Technology designers must work largely within the constraints shaped by the user organisation. Particularly for mission critical software, issues of work organisations are embedded in the more obvious issues of software design. Organisations generally function according to a dual set of realities: the formally prescribed procedures and policies and the actual practices its members use to conduct their work. These two realities are important and have implications for the effective design of software (see Weltz, 1991, on implications for software implementation).

The formal reality of organisations has been characterised as "Myth and Ceremony" by Meyer and Rowan (1991). Organisations have practices and procedures (the myths) that represent prevailing concepts of rationality but those in organisations do not actually follow them except "ceremonially." Organisations maintain this formal reality as a means of increasing "their legitimacy and their survival prospects, independent of the immediate efficacy of the acquired practices and procedures" (Meyer and Rowan, 1991, p. 41).

Formal structures of organisations develop in accordance with traditionally accepted principles of bureaucratic functioning such as a very detailed division of labour, formalised standard operating procedures that specify every allowable contingency, rigid and hierarchical lines of authority and responsibility, and objectives of minimising necessary skills at the lowest levels of the organisation.[36] Generally, powerful actors in the organisation believe these principles are the proper if not most efficient way to run and control the organisation and they provide a patina of legitimacy to the organisation to others (e.g., stockholders, board of directors, the public). There is an expectation by those inside and outside the organisation that there will be suitable mechanisms for accountability and means of dealing with uncertainty.

The formal rules that are of significant consequence for software design are those that tend to be developed as highly prescribed procedures for dealing with every contingency and providing little discretion at the lowest levels of the organisation. These rules for accountability often require not only specification of procedures for operation but also procedures that conform to accepted, generally quantifiable/measurable types of criteria.

A bank, for example, might have trouble justifying a policy of providing tellers a great deal of latitude in decision making. Many both within and without the organisation (e.g., customers, stockholders) might look askance at a bank that allowed, as policy, its lowest level employees, tellers, to have significant decision making authority. Instead, sets of strict procedures and oversight provide demonstrable accountability and responsibility to those who might review the bank's operation. It might be possible for the bank to justify decision making responsibility by tellers if it were very selective in its hiring and provided substantial training to its tellers. This, however, generally is seen as a less legitimate means to run this type of organisation and requires more resources (for salary and training) than the bank might be able to justify, that is, given the norms within the organisation and throughout the industry. In prevailing organisational norms there tends to be greater legitimacy for technology investments, particularly if they promise to enforce procedures and increase productivity according to traditional quantifiable measures (versus, for example, quality and flexibility).[37] In this regard, the thrust of formal policies and thus principles of technology design may be characterised as trying to remove control from those workers at the lowest levels who are responsible for actually producing a good or delivering a service.[38]

The technology use or design approach in service sector work and in software design is similar to that found in research on engineering of manufacturing technology.[39] Work organisation in services is often being pushed to become more like factory work, to develop an assembly line for bureaucracies. In bureaucratic organisation of service work, formal policies are developed based on the goal of eliminating discretionary action. It is assumed not just that there are known methods for achieving the goals of the organisation, but that nearly all the necessary methods, situations, and actions can be anticipated and specified. Control over individual discretion is part of the accepted means of ensuring fairness, legitimacy, and uniformity of outcomes.

Despite grand and ever-present promises, the theoretical possibilities of engineering fully automated technology controlled systems are somewhat elusive in the day-to-day reality of work. The problem is that actual operating conditions, whether a piece of physical material to be machined or a person requesting a service, regularly deviate from the specified and anticipated procedures. Even in basic manufacturing processes it is difficult to operate a fully automatic process: Metal may vary in its properties (e.g., hardness, purity) and machine tools wear or operate with some variability. Moreover, new technology as well as changes in the organisation's environment can lead to changes in the organisation and/or its processes, such as new products, new services, new methods of conducting business. These changes can introduce new conditions that were not anticipated in the technology design and may require adaptation by the people using the technology. A highly integrated, fully automated system may inhibit innovation.[40]

The nature of most service sector work makes this problem even more complicated. People's actual and perceived situations, and thus their service needs, are almost infinitely variable and thus quite frequently fall outside normal operating procedures. The ability to anticipate and plan for every contingency is limited, and dealing with it efficiently requires a response that may not be in accord with the specified procedures.[41]

The solution to this problem in many organisations has been found in "decoupling" the formal policies and the informal procedures or in designing loose integration of automated systems.[42] The demand for legitimacy and control, however, places limits on the degree of loose coupling usually permitted in organisations. One approach is found in professional occupations where norms are established but individual practitioners are expected to exercise judgement to achieve a given outcome. Education, training, and certification of workers provides legitimacy to their procedures without requirements for strict procedural observance or monitoring. In a study of hospital systems (Salzman and

Rosenthal, 1994; Chapter 6), the prevailing ethos of professionalism was used to justify soft controls in the system that allowed for decoupling, that is, of not taking advantage of the potential of software to enforce agreed upon procedures. In field service systems, however, (Salzman and Rosenthal, 1994; Chapter 5) there were two different approaches. One company did not use the system to enforce procedures because it viewed the field engineers as skilled workers who would use their best judgement and skills to perform their work properly. Assessment of the field engineers' performance was based on the final outcomes of their work (e.g., customer satisfaction) rather than adherence to detailed procedures. In contrast, the other field service company tried to use the software to enforce detailed procedures regulating the work process. The bank similarly designed the software to enforce their formal procedures. However, the implementation of the system differed in the second field service firm and the bank: When the field engineers subverted the system, the field service management and system designers responded by trying to tighten the controls to ensure compliance. In the bank, branch managers decoupled the system by allowing tellers to work around and subvert the controls designed into the system.

When legitimacy and accountability are not achieved through professionalism, the formal and informal procedures tend to be decoupled in a nonsanctioned way. People who perform the actual production or service delivery tasks are then permitted to do what is necessary only by their immediate supervisors or are able to do so surreptitiously. Just as religious ceremonies are observed with varying degrees of orthodoxy, so too the ceremony of the formal organisation is followed in varying degrees. Some in the organisation religiously try to enforce the procedures while others require only perfunctory observance.

Software systems are designed, however, according to the formal procedures and policies of the organisation. In part this is done because the software, like the organisation, gains in credibility and legitimacy if it appears to operate according to the formal policies. Mission critical software exacerbates the uneasy tension between those who enable decoupling within the organisation and those trying to constrain or restrict it. Current approaches to software design increase that tension. Change in software design and the organisation presents a formidable challenge.

Assimilation, accommodation, and the prospects for change

Technology designs may be characterised in terms of the degree of technical change as compared to previous technologies. Consider, for example, the development of word processors in which there were significantly different design approaches taken in the early models. The dedicated word processors tried to replicate the features of typing on a typewriter. The system was designed to make typing faster and easier without changes in previous routines or concepts of the tasks. This made it easier for the transition from typewriters to word processors. The technology itself might be considered an incremental advance over typewriters. In the banking case the design of the teller system was similarly oriented to appear as an incremental advance over previous systems by replicating the design of existing screens even though they were considered inefficient. In one instance the account numbers were entered as two fields, as they had been in the previous system because of limitations in the technology. The designers originally developed the system to take account numbers as one field which they saw as overcoming a limitation of the earlier systems. Users insisted that the system be redesigned to have two fields because that was they way they were accustomed to entering the account numbers. One designer characterised the request of users to have two-part account numbers as "limitations turned features." It was an effort to assimilate technology into existing work routines, to make the technology reflect existing practices and gain high user acceptance.

In contrast, the development of modern word processing systems on standalone personal computers can be seen as radical advances in technology when compared to the earlier systems. These systems do not try to preserve "typing" but rather transform the task into document composition. Documents appear to scroll continuously rather than be formatted page-by-page. Additional text is entered and footnotes added or deleted without any page adjustments by the user. The attributes of the technology could be characterised as a radical advance over typing by embodying a vision of the work task as document composition rather than typing.

The characterisation of the technology on a continuum between incremental and radical represents attributes of the technology as the designers constructed it. "User requirements" are identified and interpreted through the filter of the organisation of user and vendor firms and their environments and embodied as particular physical and functional characteristics of the technology. The technology, in short, reflects an interpretation or vision that was shaped by the structure of the design process. User organisation attributes and system requirements are first expressed and then interpreted through the structure of the user organisation, the market, and the vendor organisation. These attributes and requirements are then interpreted by the designers, within additional constraints of the design space, as they develop the technology.

When the technology is adopted by an organisation, it becomes a constraining, though not determining, structure that in part shapes the action space of its user. The technology in the user organisation interacts with other organisation features. Technology implementation and use is a process of reinterpretation and redesign according to organisational characteristics. Here, the technology as designed confronts the requirements of the informal organisation and the conflicting, often irresolvable interests of different users, and brings to existing routines new technological possibilities and constraints that, in part, redefine the action space of the user. This leads to a reinterpretation of the potential of the technology, in effect a change from its intended design. Technology is not received by an organisation as it was originally defined in part because of the inherent differences between the process of defining and of using technology.

What does it mean to "adopt" a new technology? On the one hand the user organisation might assimilate the technology to fit into existing structures and modes of operation. Alternatively, it might change its work processes to accommodate new features and functionality of the technology. Which of these two situations actually dominates will be influenced by the attributes of the technology in combination with an organisation's characteristics such as its general approach to organising the work process.

Using the example of a word processor, we can construct a matrix of technology attributes and implementation and use characteristics (see Figure 5.2). For this example, the word processor can be designed as an incremental change over a typewriter or as a radical change. Each can be assimilated by the user organisation or lead to reorganisation of the work process as an organisational accommodation of the technology. The early word processors could be assimilated to replace typewriters and both the work task and organisation of work might remain largely unchanged. However, the higher cost of word processors might lead some organisations to increase their utilisation through creation of word processing pools with specialised word processing secretaries. The latter reorganisation would be an accommodation of a technology that was designed to be only an incremental change over typewriters.

A technology designed to be a radical innovation, such as a personal computer for word processing, could also be implemented and used by the organisation through assimilation or accommodation. The organisation assimilating it might use the computer as just a faster typewriter and not substantially change the work process in terms of the substance of the work or responsibilities or skills of the users. (Its lower cost than the original dedicated minicomputer word processors might allow greater assimilation because the organisation could afford to replace typewriters on every desk.)

An organisational accommodation might change the substance of the work, perhaps having the authors of letters and documents type the first drafts and assigning more sophisticated graphics, spreadsheet, and document formatting tasks to secretaries. In an even more radical change in the organisation, as occurred in one large computer company, secretaries no longer did any typing for others as everyone was given a personal computer and printer and expected to do all his or her own typing.

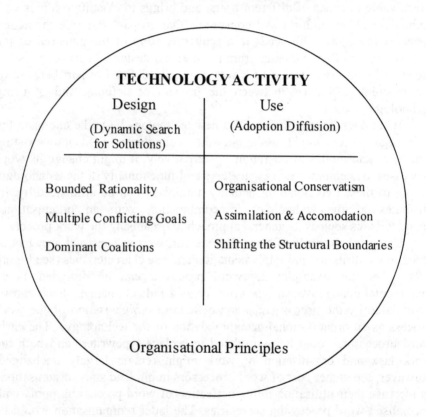

Figure 5.2   Organisational principles underlying technology design and use

Something new can be assimilated in ways that shore up existing routines and power arrangements, but it can also be assimilated in ways that provide opening for change if inconsistencies and contradictions develop. Paint has shown how a child will assimilate and ignore contradictory evidence until developmentally ready to develop a new schema or structure. In our model, it is through a social process (rather than biological or developmental) that a combination of factors, technology, organisation, market environment, or the nature of particular people within the organisation strain the existing structures (cf. Kuhn's, 1970 analysis of paradigm shifts). A failure to assimilate fully new conditions and technology will strain the existing structures. It can be incremental and continuous or radical and discontinuous. Often stasis can be maintained for long periods, with the organisation resisting change even in the face of continued problems, until some traumatic event becomes a crisis and unfreezes existing structures, allowing change to occur.

Software is potentially a technology that can precipitate such crises by imposing formal procedures and changing the collection and uses of information. This kind of technology therefore becomes increasingly difficult to assimilate into existing structures. In this way, the expansive power of new computer technologies can lead, in unintended ways, to organisational changes that, in turn, allow for better design and use of the technology.

The coming challenge of design

As the medium through which work tasks are performed, software provides the means to implement and regulate procedures to a degree not possible in noncomputerised bureaucracies. When user requirements are defined only as the formal policies of the organisation, there is little latitude for the informal practices to be sanctioned by those designing the software. Under such constraints, systems that were designed to embody informal work practices would lack legitimacy. Thus, if software that reflects the formal policies is implemented as designed, it would collapse the divergence between the formal and informal realities of the organisation. That is, because software mediates the actual execution of work, because it operates as a tangible technology, there is much less negotiation possible than when formal policies of the organisation are embodied as operating procedures that depend upon people for their implementation. There is less space or latitude for the informal realities of the organisation to coexist with the formal structures of the organisation if the software system is implemented as designed. Thus, we can see a source of dysfunctionality of software as coming from structural or inherent features of organisations rather than from an inadequate technical process of user requirements determination.

One design solution is to use "soft" controls in the software. This design option often depends upon the legitimacy of the users as autonomous decision makers and their power to resist encroachment on their territory. Soft controls in software are generally possible only in organisations with intentional explicit loose coupling and accountability through professional norms. In most other cases, however, software that does not reflect formal policies and procedures may be difficult to sell to user organisations. The "buyer" and those in upper levels of management, who generally endorse the formal organisation, exert the most influence in defining the systems and in procurement decisions.

Incorporating traditional types of controls in the software design often leads users to subvert aspects of its intended functionality. In a bank, for example, tellers and their supervisors found ways to defeat the security provisions built into the system. In a field service system, the field engineers devised ways to "pencil whip" the data to make the monitoring function ineffective. Sometimes these software functions are just regarded as poor design features or as a burden to endure, as a ritual that the organisation requires and to which workers need to pay homage. If any thought is given to these features or functions of the software, they may be seen as serving a legitimating function for the organisation (e.g., providing data for tracking productivity, however suspect the data may be, or providing at least appearances of having approved procedures that regulate the work process). Others, however, will believe in enforcing the prescribed procedures and work to increase the hard controls in the system and devise ways to identify users who do not conform. Much like battles over Victorian morality, there are the true believers who fear the dissolution of society without at least formal observance of prevailing norms and there are those who appear to capitulate by dutifully agreeing yet follow their desires into unsanctioned behaviour.

When software systems are designed in accordance with the formal reality of organisations with hierarchical decision making and authority, the outcome is generally a system that must be subverted to restore flexibility to end users or that leads to an organisation that hobbles along with a dysfunctional system. In some environments firms can survive quite well, or at least survive for long periods of time, with systems that do not provide the necessary flexibility to end users. In dynamic environments, with high pressure for efficiency and competitiveness, there is less latitude for organisations to continue with dysfunctional operations. Changes in the competitive environment of organisations -- especially recent changes in industries facing global competition and increased emphasis on quality over sheer productivity -- may increase the tension in those organisations trying to implement formal policies and procedures through software.

This issue is not solved simply through improved methods for identifying user requirements. Instead it calls for a more complex process of generating fundamentally new operating alternatives within the user organisation. Alternative forms of work organisation, for example, allow for developing new means of accountability. Some firms are experimenting with new forms of work organisation that expand the role of frontline or shopfloor workers. The alternative scenarios require the simultaneous and deliberative restructuring of software design and the user organisation in ways that are mutually compatible. The vendor alone cannot create new technology. There needs to be a redefinition of the design space and the user space. This involves changes at a number of levels. It requires changes in the cognitive maps of designers but they must be sustained, or motivated, by larger changes in the organisation norms and the external environment such as market demand. Market changes involve the user organisation which similarly must change at a number of levels. This includes not only management philosophy but often control over resources and power in the organisation. There generally needs to be a more sustained and extensive pressure on the user organisation to change. The success of new technology, such as new software designs, requires changes in the user organisation to accommodate it rather than assimilate it into existing patterns or operating procedures.

Ironically, technology is often an object of contention because it can provide an opportunity or catalyst for change and thus, in practice, its potential to spark change is likely to be muted. Traditional user requirements analysis will tend to be dominated by those in the organisation with more power who will generally define the "objective requirements" as those that maintain the status quo and specifically their own position. However, the software, because it can implement the formal reality of the organisation in an uncompromising way, may lead people in the organisation to confront the conflict in the dual reality of the organisation and, through this confrontation, provide an opportunity for change. If this opportunity is taken, the outcome can lead to new ways of maintaining organisation legitimacy through, for example, a professional model of responsibility and accountability for frontline service workers. However, the change opportunity must be broader than merely designing new types of technology because the software designer alone cannot reconcile the user organisation's formal and informal visions of needed functionality.

Ultimately, the solution to this aspect of the problem of software design is to be found in the solution to the problems of how the work process is organised. Depending on an organisation's particular service mission, customer satisfaction requires a blend of consistency and responsiveness in the process of service delivery. Accordingly, to realise the full potential of mission critical software, organisations must avoid becoming so entrenched in maintaining the facade of formal policies and procedures that they lose the ability for effective, flexible work performance. In this regard, one important change in the user organisation that has implications for software design is acceptance of soft controls rather than trying to use the theoretical potential of software to implement polices completely through hard controls.   Taking this step requires reconceptualising the role of the worker in the work process as a knowledgeable, responsible, and skilled person and of the workplace as best operating with some degree of discretion.   It also requires changes in the traditional conceptions of technology and giving up on the engineer's common dream of full automation.   In short, it means looking for effective organisational solutions rather than narrow technology fixes. It means that software becomes more a tool for the user than a means for control.

Notes

1.   This point was first made by Agneta Olerup (1993) in her review of an early version of our book, Software By Design.   She concludes that, despite the problems with the term "user," there is no good alternative.

2.   In the late 1980s a coherent body of work on the social shaping of technology began to appear.   This is represented in collections such as Bijker, Hughes, Pinch and Trevor (1987) and Dierkes and Hoffmann (1992).   The former have brought together the sociology and history of technology into an area labelled Social Construction of Technology (SCOT), some building on perspectives in the sociology of science (e.g., Woolgar and Grint, 1991), and the Science, Technology, and Society (STS) field.   Much of this research does not specifically address workplace technology.   Some of the process technology innovation research, which does concern workplace technology, is more focused on the process of innovation, of the factors shaping the inception of the technology, initial design choices, and the diffusion trajectories rather than on the interaction of workplace structure and values and the design of specific technology.   Recent research that does address the social side of workplace technology design includes Brîdner (1990), Corbett, Rassmussen, and Rauner (1991), and Ehn (1989), among others.

3.   The work in the general area of social construction of technology has been an important counterbalance to the long tradition of studies that abstract technology from its social context of creation and use.   This approach is important for focusing attention on how technologies are constructed and used through interpretation by users. That is, how the "meanings" of technology are constructed by users rather than inherent to a technology. These approaches provide important analyses of the shaping of technology within the user space. Some argue that technology is just one type of node in actor networks and, in this network, technology holds no special status or attributes vis-à-vis other parts of the network. That is, the people/technology distinctions are not particularly salient.   Some of the writing in the extreme social constructivist school seemingly argues that there are no salient properties of technology because all depends on social context and interpretation (e.g., Woolgar and

Grint, 1991; Grint and Woolgar, 1992; Kling, 1991, 1992, for a rejoinder). Certainly the variability and inconsistency of perceptions about technology and its impacts suggest that technology is not absolutely determinant in its impact. However, at times it does seem that the background "fuzziness" and contingency of propositions about technology and its impact become the dominant foreground for the social constructivist analyses. Instead, we argue that interpretations of technology by individuals are bounded and structured in some systematic ways that allow for analysis at the supraindividual level. In a review of German perspectives on social shaping of technology, Rammert (1992, p. 83) writes: The works based on the social constructivist approach have had a mixed reception among German researchers of technology. On the one hand, they have been faulted for what is perceived to be their exaggerated theoretical pretensions. The interactional level is said to be treated as absolute at the cost of the sociostructural level....On the other hand, the empirical studies and the related procedures of the social constructivist approach have encouraged proponents to expand the field of inquiry and increase their consideration of sociocultural influences.

4. The process and the language analogy are based on a framework developed in structuration theory and interpretations of the role of language and the central role of power influencing design decisions and the "interpretations" by actors (e.g. Giddens, 1984, 1991; Livesay, 1989).

5. The term values is used in a broad sense to refer to the explicit and implicit, the acknowledged and unacknowledged beliefs, assumptions, and judgements that inform and shape people's actions and provide the schemes with which they interpret the environment. It shapes their views not only of how the world is, but also of how the world should be, providing a normative framework in addition to an interpretative one. At the same time, it is important to examine the structures of organisation and the social environment that shape people's values. People's values will differ, in part, as a result of, for example, their position in the organisation. The position of manager may tend to foster different values than position of engineer. It is also through "structure" that values are perpetuated. The role of values as shaping and as being shaped by structures can be viewed in the context of structuration theory (e.g., Giddens, 1985, 1991; Livesay, 1989; see also, Ortmann, Windeler, Becker, and Schulz, 1990, for a broader discussion of these issues in technology and organisation). Sewell, 1992; Becker, Windeler, and Ortmann, 1989; Orlikowsky, 1992; Barley, 1986; Bourdieu, 1991).

6. Giddens has developed this theory through a number of books (e.g. 1984, 1991) and others such as Sewell (1992) and Livesay (1989) have addressed specific dimensions of the theory. It has been discussed in terms of technology and organisations by Becker, Windeler, and Ortmann (1989), Orlikowsky (1992), and Barley (1986).

7. See, for example, Libenstein (1987) on the inefficiency of organisations. It is also apparent in the survival of many very large organisations, such as General Motors, with very inefficient organisations.

8. Most structural changes have led to some flattening of the hierarchy, removing middle levels in the chain of command and delegating decision making to lower levels. This may widen the scope of decision making by lower level workers, for example, but generally does not change a structure of hierarchical power distribution. At some point the degree of power and decision making by lower level workers could become quite large, eliminating many of the traditional hierarchical power issues (e.g., decisions about work process, work rules, etc.). However, evidence to date suggests that most firms have not significantly flattened their hierarchies, expanded the power of lower level workers or implemented in any significant manner work reorganisation. In a survey by the National Center on Education and the Economy (1990, p. 16) it was found that "95 percent of United States companies still cling to old forms of work organisation." Similar results were found in a study of worker involvement in technology design and production organisation in manufacturing (Salzman, 1992; Lund, Bishop, Newman, and Salzman, 1993). Fortune magazine (Kiechel, 1992) observed that "In the course of the current economic unpleasantness progressive management has taken it on the chin...When business got bad, the top brass at too many companies sucked all the decision-making back to the top." Russell's 91988, p. 374) review of employee participation in the U.S. finds that, although there is widespread discussion of various forms of quality circles, work redesign, and other programs, "research on the impact of these programs, however, suggests that they tend to have either negligible or at best only short-lived effects. And American managers continue to shy away from the major commitments that appear needed to increase the likelihood that these efforts to increase employee participation will have both significant and lasting effects."

9. An interesting case has been made by David (1985) that, basically, historical accident in combination with other social factors can lead to a "lock-in" of less profitable, less productive technologies. In this essay on the standardisation of the typewriter keyboard layout (known as the QWERTY keyboard), David argues that in the early days of the typewriter the QWERTY keyboard had a slight edge over other layouts. Although this keyboard was less productive than other designs, typists were beginning to be trained in its use and thus came the expectation that most typists would continue to be trained that way. With high turnover, employers decided that it was unprofitable for any individual employer to retrain his or her employees. It was this early lock-in that quickly led to widespread standardisation of the QWERTY layout. David (1985, p. 336) writes, "despite the presence of the sort of externalities that standard static analysis tells us would interfere with the achievement of the socially optimal degree of system compatibility, competition in the absence of perfect futures markets drove the industry prematurely into standardisation on the wrong system where decentralised decision making subsequently has sufficed to hold it" (emphasis in original). Relevant to our argument is this example and others that design can be socially shaped even in the face of economic disincentives, or at least to recognise that the economics of profitability may relate to a network of social choices.

10. They identify a number of factors about organisations that inhibit innovation such as the resistance to change from organisational actors who are invested in the existing structures of power, existing labour structures that may not accommodate new job arrangements, and that few incentives may be provided for change (Child, et al., 1987, p. 111).

11. As Paint (1968, p. 63) describes the process, "assimilation is the process whereby an action is actively reproduced and comes to incorporate new objects into itself, and accommodation, the process whereby the schemes of assimilation themselves become modified in being applied to a diversity of objects." Piaget's analysis of cognition relies on a developmental/biological model, whereas our analysis is based on this process as a social process. It is the process of structuration that frames the interaction between technology and organisation.

12. This can also be seen in the design of earlier word processors such as the Wang system. Dedicated word processing software for use on minicomputers tried to replicate manual typing procedures in order to ease the transition from typewriting to word processing and increase user acceptance. (One company used black type on a white background, advertising the similarity to using a typewriter and white paper.) These systems computerised typing tasks rather than computerising document composition. They allowed the systems to be assimilated into existing work processes but they did not utilise the technical capabilities of computers to transform the procedures of document composition. However, in the case of the minicomputer-based word processors, preserving existing procedures did not always result in preserving working conditions, a variety of changes occurred, some of which preserved some aspects of working conditions while changing others. Minicomputer-dedicated word processors were sometimes viewed as too expensive to directly replace typewriters and were seen as an opportunity to rationalise secretarial work. In these instances, secretarial work was fragmented and word processing pools were created, exclusively devoted to typing. In contrast, the use of microcomputers to replace typewriters preserved previous working arrangements while hosting word processing software that transformed the work procedures.

13. There are several problems with the term "user." From the software developer's perspective the term provides a generic label for people using the computer and there are many common aspects of computer use for all users. However, it is a problematic term for at least two reasons: First, the term tends to obscure the identities, roles, and functions of specific users in their jobs as bank tellers, accountants, sales managers, etc. Computer use becomes a common denominator that takes on greater importance than the specific application. Second, and perhaps more importantly, it emphasises the importance of technology over people in which users are just adjuncts to the computer, thus inverting the relationship between people and computers. Instead, it should be recognised that people use the computer as a means of accomplishing their tasks. Related to this, and our point in this section, is that the term "user" decontextualises people from their roles as workers. Agneta Olerup (1993)

identified this issue in an earlier version of this manuscript and we agree with her in both these observations and in her conclusion that there is no good alternative to the term "user" for these discussions. Writers and additional tasks such as spreadsheet tasks and graphics preparation are being performed by secretaries (See Johnson and Rice, 1987, chapter 4, for further discussion of different word processing adoption approaches).

14. Those who are not fused and accepting of the common goals of the organisation are "the confused, frustrated, and defeated [who] all suffer a common malady. They have not responded to the socialising process successfully...Lack of ability or lack of adaptation may have caused their problems" (Miller and Form, 1980, pp. 134-135). In Miller and Form's view, seeking satisfaction from work is futile for most people. Workers "are still encouraged to seek the values denied by the technical and economic realities of life. Failing to fulfil their social wants, they blame their failure on the organisation rather than on the sociocultural orientation that leads them to look for security, satisfaction, and status, where they are in short supply." It is in "the approximately half the hours in each week for activities other than working or sleeping...[that] workers may exercise their options to engage in recreation, eating, participation in voluntary associations, sex, sheer idleness, and the like....Motivation generated in and toward out-of-work life can supply the will to work in factory or office." They go on to discuss, somewhat dismayed, that cultural changes in society make younger workers "much less willing to accept authoritarian influences at work or in the community" (pp. 154-155). Although the form of expression in these passages may seem a bit dated, it still represents a predominant underlying perspective about work organisation and work: The issue about whether, and to what degree, work should be a source of satisfaction is still debated. (This theme runs from early sociology of work such as Dubin (1956), to current social theorists such as Offe (1985) and Gorz (1980)).

15. One of the notable articles is that by Meyer and Rowan (1991) and also discussed by Perrow (1984, 1986) and by Weltz (1991) specifically in terms of software. It is also discussed in studies of the workplace such as in Hirschhorn's (1984) work.

16. For example, Noble (1977, 1984), Gordon, Edwards, and Reich (1982) and Marglin (1974).

17. See Salzman (1991, 1992) and Lund, Bishop, Newman, and Salzman (1993) for review of hardware design approaches.

18. From Computer News (1986), cited in Friedman (1989, p. 171).

19. The effectiveness of customer involvement in product design is discussed in Rosenthal (1992) and the equivocal findings about its effectiveness in software are discussed in Ives and Olsen (1984), King and Rodriguez (1981), Tait and Vessey (1988), and Friedman (1989). For other perspectives on limitations of user involvement, and how some of these problems can be addressed, see Robey and Farrow (1982), Franz and Robey (1984), Boland and Day (1989), Newman and Noble (1990), and Grudin (1991a, 1991b).

20. These books were all the textbooks being used in computer science and Management Information Systems courses at Boston University and a random selection of textbooks in the library. For overview and discussion of the technical and general problems of software development, see Cusamano (1991) and Friedman (1989).

21. For comprehensive overviews and reviews, see Attewell and Rule (1984), Kling (1980), and Friedman (1989).

22. A few representative studies are: Kling (1984), Bariff and Galbraith (1978), Markus (1984), Markus and Pfeffer (1983), Robey and Ferron (1982), and Boland and Day (1989).

23. See Markus (1984), Markus and Pfeffer (1983), Kling (1985), Danziger and Kraemer (1986).

24. In the U.S. political issues tend to be framed in terms of special interests of groups and coalitions representing pluralistic interests. There is a strong belief in mobility and class politics are not explicitly discussed nor is class position seen as an enduring characteristic. The perspective in the U.S. is shaped by experiences and folklore of a greater mobility in the U.S. than in other countries. For example, although the Democratic party usually represents organised labour, it is not a "Labour Party." In the U.S. there is a decline in the perception that there are particular interests shared by workers. This is reflected in declines in unionisation and a declining influence and distinguishable political agenda by labour unions. This is a sharp difference when compared to the influence in various parts of Europe and particularly in Scandinavian countries and it is reflected in differences in work organisation, including technology policies.

25. Although influenced by U.S. human relations theory, sociotechnical researchers maintain a distinct approach, according to Friedman (1989, p. 196), because "they believe that once conflict has been built into a work situation only structural change can remove it." The sociotechnical approach focuses on resolving conflict between management and workers by emphasising "their common interests in the development of new technology, which is based on their mutual interests in preserving workplaces" (Bansler, 1989). Sociotechnical approaches tend to involve deeper organisational change than the U.S. Quality of Worklife and other programs. One exception in the U.S. is Pava (1983), one of the few researchers to address.

26. This is due, according to Bansler, to the "idealistic and inadequate understanding of the driving forces behind the technological transformation of work." Bansler's (1989, p. 15) critique is that the Scandinavian approaches (or, more accurately, the "critical tradition") reject "the harmonious view of social relations in the workplace, which imbue the systems theoretical and socio-technical research traditions." Instead, organisations should be viewed as "frameworks for conflicts among various interest groups with unequal power and resources. Social relations at work are characterised not only by Cupertino but also by conflicts and struggles between managers and employees, and among different groups of employees...Since these conflicts are related to the class structure of the society, they cannot be 'resolved' or abolished at the level

of an individual organisation, as the socio-technical tradition believes. The 'participative approach' to systems development is not sufficient." Instead there needs to be a broader based change in the structure of work systems, if not in society. Sociotechnical principles in office work who outlines a strategy that requires significant organisational restructuring to make best use of new technology and create satisfying jobs.

27. One study of Canadian and Danish software designers' values (Kumar and Bjorn-Andersen, 1990) found systematic and significant differences. They found that although there was a dominance of technical and economic values (versus sociopolitical values such as concern about organisational and job satisfaction) in both countries, Danish designers were more concerned with these social issues than the Canadian designers. Other examples of the differences in industrial culture and technology may be seen in the case of Computer-Aided Design (Majchrzak and Salzman, 1989).

28. The concept of industrial culture as used here has been developed by Rauner and Ruth (1989).

29. For example, the Scandinavian approaches can provide a model for ways to reorient design but also show that their particular approach reflects a particular industrial culture shaped by, among other factors, a longstanding craft tradition, a workplace environment of 90 percent unionisation, and requirements for labour union participation in many basic decisions that would be considered management prerogatives in the U.S.

30. Although discrete entity consensus model studies may consider the social aspects of the developer-user interaction, remedies are cast as improvements in methods, techniques, and intentions of system designers and users. Power and values do not enter into such discussions because consensus models of social systems do not recognise inherent or structural value conflicts as legitimate, that is, intrinsic to existing organisations and social systems.

31. In a review of the research and literature, Kling (1986, p.3) notes that he did not identify any web models in the computer science textbooks. He found the predominant perspective of computers was one of discrete entities, unrelated to broader organisational dynamics.

32. Structuration theory offers some important possibilities for a new model of analysis of technology and organisations that, we find, is most important in viewing this dynamic.

33. Organisations attempting to "empower" the workforce, a common theme in U.S. management in the 1990s, are not immune to this constraint. Empowerment, typically seen as a way of seeking incremental improvement in organisational performance, frequently stops short of playing a key role in the design of new technology. See Chapter 2 for further discussion of this point. Technology-organisation relationship. We have found such theory generally useful in our analysis of service delivery cases involving new software. Thomas' (1993) research addresses similar issues in the manufacturing environment. Thomas (1993, Chapter 6) adds some additional elements to structuration theory to develop what he refers to as the "power- process" approach. He find consideration of "purpose" lacking in structuration theory

and argues that: in order to understand how and why exogenous developments come to be recognised or how and why strategic choices come to be made, it is essential to understand who or what segments of an organisation have the power with which to define the parameters of search for problems and solutions, the criteria with which alternatives are evaluated, and the manner in which a choice once made is implemented. Of necessity, the power-process perspective accords a central role to power relations in the organisation. We agree with Thomas' emphasis on these particular theoretical elaborations.

34. Robey and Markus (1984, p. 12) show how defining system requirements is a political process, at least in part, and how information requirements analysis is distorted, serves legitimating functions, and is conservative of the status quo: Regardless of whether it actually produces rational outcomes or not, systems development must symbolise rationality and signify that the actions taken are not arbitrary, but rather acceptable within the organisation's ideology. As such, rituals help provide meaning to the actions taken within an organisation [emphasis in original]. While showing the ritual/political part of systems development, they stop short of analysing any systematic tendencies or content issues versus process at the individual level. They do not discuss implications for design beyond "beware of what is really going on" (p. 13) in the process.

35. The implications for software design can be seen in the example of design choices for implementing controls (e.g., hard versus soft controls). Robey and Markus (1984) and other analyses on software design locate the locus of change with the designers. This is limited because designers have limited latitude in which to manoeuvre. Those designing the technology do not, and cannot, act with great freedom from their social environment. Their design space is constrained by factors beyond their own perceptions and perspectives. Thus, the focus of design changes needs to be much broader than the attitudes, personalities, and education of engineers. We need to view their constraints as also related to an environment of organisations, markets, and industrial culture of the society. The term industrial culture is used to denote the nontechnical factors that shape technology and that can be examined as related to specific societies. That is, thinking in terms of industrial culture.

36. As Meyer and Rowan (1991, p. 45) explain, the operating procedures for organisations are prefabricated formulas available for use by any given organisation...Similarly, technologies are institutionalised and become myths binding on organisations. Technical procedures of production, accounting, personnel selection, or data processing become taken-for-granted means to accomplish organisational ends. Meyer and Rowan (1991, p. 45) also note that "Quite apart from their possible efficiency, such institutionalised techniques establish an organisation as appropriate, rational, and modern. Their use displays responsibility and avoids claims of negligence." Not only, for example, do formal accounting procedures and economic analysis provide legitimacy and rationale for a decision, "such analyses can also provide rational accountings after failures occur: managers whose plans have failed can demonstrate to investors, stockholders, and superiors that procedures were prudent and that decisions were by rational means" (Meyer and Rowan, 1991,

p. 51) suggests how different industrial countries deal with issues of technology design and production organisation. Rauner and Ruth (1990, p. 121) list important dimensions of industrial culture as "national traditions and societal institutions; organisational preferences; social institutions and government policies; educational institutions; social (and individual) psychology." Thomas (1993, Chapter 1) also proposes that the politics of technology are important to understanding technological choice and use, arguing that we should: conceive of the relationship between technology and organisation as mediated by the exercise of power, i.e., by a system of authority and domination that asserts the primacy of one understanding of the physical world, one prescription for social organisation, over others...[it is] the opportunity...for different categories of organisational actors to try and put in place their own unique world views about the "proper" way to organise work [emphasis in original]. In earlier studies of engineering and technology design, Perrow (1983, 1986, p. 151) similarly concluded that "in general equipment and technology are chosen to require and reinforce centralised authority structures, even where decentralised ones would be more appropriate."

37. This point has been made by many researchers and was found to be an important part of technology justification in studies of manufacturing technology design (e.g., Salzman and Lund, 1993; Lund, et al, 1993), and was explored in detail by Thomas' (1993, Chapter 6) study of manufacturing technology where he found that "the idea of technology, and, more specifically, the idea of automation as a labour-saving device, can itself become institutionalised...the presumed benefits of automation have become so taken-for-granted that in the absence of overwhelming proof that other alternatives are possible, automation becomes the default option." It is not only the value of technology solutions as an ideology but also the means of justification that support this ideology: "...the subordination of process to product helps explain why manufacturing managers and engineers would adhere to traditional return on investment metrics. In most cases, those metrics provided a functional substitute for an explicit manufacturing strategy." When there was uncertainty,"

38. The goal of engineering is to embed mechanisms for control over the worker and the process in the technology. In this approach there tends to be a devaluation of people and their potential contribution to the work process. David Noble (1984) observed that engineers traditionally express a "a delight in remote control and an enchantment with the notion of machines without men...a general devaluation of human skills and a distrust of human workers and an ongoing effort to eliminate both." As in bureaucratic procedures, in technology design there is a belief that all contingencies can be anticipated and processes can be completely and continually controlled through specified procedures. Similar findings are reported in Perrow (1983) and in Thomas (1993). This perspective in engineering was found in our survey of software design and engineering textbooks and reflected in other interviews with engineering managers (see findings as reported in Salzman, 1992; and Lund, et al., 1993). As noted in this review, although management theory has evolved

during the post-war period, design principles do not appear to have undergone a similar development. The review of design textbooks indicated that engineering proceeds with the same basic understanding of the human role in production as first articulated at the beginning of the twentieth century....the default option [i.e., traditional return on investment metrics] was also the safest: to restrict the search for both problems and solutions that fit with traditional measures even when doing so might produce deleterious consequences (e.g., increases in the volume of indirect labour). On those few occasions when managers and engineers chose to go out on a limb, they ardently resisted arguing for alternative measures because having the numbers even numbers that they might have ridiculed in private enabled them to argue that their choices were legitimate. The numbers were legitimate because they had been screened through a procedure that was deemed to be legitimate and defensible.

39. Although much of the software design literature is more advanced in its recognition of the cognitive dimension of humans and the role of users in design, it does not depart substantially from a basic engineering paradigm, as discussed in Chapter 3. The cognitive psychology models used in systems design, for example, have been criticised for being mechanistic and deterministic models of human functions (Coulter, 1979, 1983; Dreyfus, 1979).

40. These problems and the general problem of automation in manufacturing are discussed in general by Noble (1984) and in the case of attempted automation of printed circuit boards using Computer-Aided Design in Salzman (1989), flexible manufacturing systems in Graham and Rosenthal (1986), and in other manufacturing cases in Lund, et al. (1993), Salzman and Lund (1993), Hirschhorn (1984), and Adler (1986).

41. As Brunsson (1985, p. 4) has observed in The Irrational Organisation: "Efficiency seldom goes hand in hand with flexibility. Coordinating different people's actions also means reducing the range of actions available to each one of them. And while the reduction in variety may increase efficiency, it also tends to undermine the ability to promote new values, to perform new tasks or to handle new situations".

42. Several studies of decoupling in organisations and in manufacturing systems are Meyer and Rowan (1978), Perrow (1984), Hirschhorn (1984). Accounts of actual shopfloor activity and their divergence from formal polices and procedures are described in Burawoy (1979) and in Hampers (1992).

References

Aitken, Hugh G. J. 1960. Taylorism at Watertown Arsenal; *Scientific Management in Action*, 1908-1915. Cambridge, MA: Harvard University Press.

Attewell, Paul and James Rule. 1984. "Computing and Organisations: What We Know and What We Don't Know." *Communications of the ACM* 27:1184-1192.

Badham, Richard and Bernard Schallock. 1991. "Human Factors in CIM Development: A Human Centred View from Europe." *International Journal of Human Factors in Manufacturing*, April, 121-141.

Bansler, Jšrgen. 1989. "Systems Development Research in Scandinavia: Three Theoretical Schools." *Scandinavian Journal of Information Systems* 1(August).

Bariff, Martin and Jay Galbraith. 1978. "Intraorganizational Power Considerations for Designing Information Systems." *Accounting Organisations and Society* 3:15-27.

Barley, Stephen R. 1986. "Technology as an Occasion for Structuring: Evidence from Observations of CT Scanners and the Social Order of Radiology Departments." *Administrative Science Quarterly* 31:78-108.

Becker, A., Windeler, A. and GŸnter Ortmann. 1989. "Computerisation and Power: A Micropolitical View." Paper presented at the *9th EGOS-Colloquium*. Berlin, 11-14 July.

Bijker, Wiebe E., Thomas P. Hughes, and Trevor J. Pinch. 1987. The Social Construction of Technological Systems: New Directions in the Sociology and History of Technology. Cambridge, MA: MIT Press.

Bittner, Egon. 1983. "Technique and the Conduct of Life." *Social Problems* 30(3, February): 249-261.

Bjorn-Andersen, Niels and Bo Hedberg. 1977. "Designing Information Systems in an Organisational Perspective." Prescriptive Models of Organisations: *Studies in the Management Sciences* 5, edited by P. C. Nystrom and W. H. Starbuck. New York: North-Holland Publishing Company.

Bjorn-Andersen, Niels and Paul Pedersen. 1980. "Computer Facilitated Changes in the Management Power Structure." *Accounting, Organisations, and Society* 5:203-217.

Blackler, Frank and Colin Brown. 1986. "Alternative Models to Guide the Design and Introduction of the New Information Technologies Into Work Organisations." *Journal of Occupational Psychology*, January, 187-313.

Boehm, Barry W. 1981. *Software Engineering Economics*. Englewood Cliffs, NJ: Prentice-Hall.

Boguslaw, Robert. 1965. *The New Utopians, a Study of System Design and Social Change*. Englewood Cliffs, NJ: Prentice-Hall.

Boland, Richard J. Jr. and Wesley F. Day. 1989. "The Experience of System Design: A Hermeneutic of Organisational Action." *Scandinavian Journal of Management* 5(2):87-104.

Bourdieu, Pierre. 1991. *Language and Symbolic Power*. Cambridge, MA: Harvard University Press.

Bowerman, Robert G. and David E. Glover. 1988. *Putting Expert Systems Into Practice*. New York: Van Nostrand Reinhold Company.

Bramel, Dana and Ronald Friend. 1981. "Hawthorne, the Myth of the Docile Worker, and Class Bias in Psychology." *American Psychologist* 36 (8, August).

Braverman, Harry. 1974. *Labour and Monopoly Capital*. New York: Monthly Review Press.

Bršdner, Peter. 1990. *The Shape of Future Technology - the Anthropocentric Alternative*. London: Springer-Verlag.

Brown, Alex and Sons, Inc. 1986. *"Computer Services Industry Overview: The Move to Mission Critical Systems."* Baltimore, Maryland.

Brunsson, Nils. 1985. *The Irrational Organisation.* New York: John Wiley and Sons.

Burawoy, Michael. 1979. Manufacturing Consent: *Changes in the Labor Process Under Monopoly Capitalism. Chicago:* University of Chicago Press.

Carey, Alex. 1967. "The Hawthorne Studies: A Radical Criticism." *American Sociological Review* 32:403-416.

Cern, Frank. 1989. *"Information Management: Study Finds Bedside Terminals Prove Their Worth."* Hospitals, 5 February, 72.

Cherns, Albert. 1987. "Principles of Sociotechnical Design Revisited." *Human Relations* 40(3):153-162.

Child, John, Hans-Dieter Ganter and Alfred Kieser. 1987. "Technological Innovation and Organisational Conservatism." *New Technology as Organisational Innovation: The Development and Diffusion of Microelectronics,* edited by Johannes M. Pennings and Arend Buitendam. Cambridge, MA: Ballinger Publishing Company.

Cockerham, Williams C. 1986. *Medical Sociology.* Englewood Cliffs, NJ: Prentice-Hall, Inc.

Coombs, R., D. Knights and H. C. Willmott. 1992. "Culture, Control and Competition; Towards a Conceptual Framework for the Study of Information Technology in Organisations." *Organisation Studies* 13(1):51-72.

Corbett, J. Martin, Lauge B. Rasmussen and Felix Rauner. 1991. *Crossing the Border: The Social Engineering Design of Computer Integrated Manufacturing Systems.* London: Springer-Verlag.

Cotton, John L., David A. Vollrath, Mark L. Lengnick-Hall and Kirk L. Froggatt. 1990. "Fact: The Form of Participation Does Matter; a Rebuttal to Leana, Locke, and Schweiger." *Academy of Management Review* 15(1):147-153.

Cotton, John L., David A. Vollrath, Kirk L. Froggatt, Mark L. Lengnick-Hall and Kenneth R. Jennings. 1988. "Employee Participation: Diverse Forms and Different Outcomes." *Academy of Management Review* 13(1):8-22.

Coulter, Jeff. 1979. *The Social Construction of Mind.* London: Macmillan.

Cusumano, Michael A. 1991. *Japan's Software Factories: A Challenge to U.S. Management.* New York: Oxford University Press.

Danziger, James N. and Kenneth L. Kraemer. 1986. "People and Computers." *The Impacts of Computing on End Users on Organisations.* New York: Columbia University Press.

David, Paul A. 1985. "Clio and the Economics of QWERTY." *The American Economic Review* 75(May): 332-337.

Davis, Louis and James Taylor. 1975. "Technology Effects on Job, Work, and Organisational Structure: A Contingency View." *The Quality of Working Life.* New York: Free Press.

Davis, William S. 1983. *Systems Analysis and Design: A Structured Approach.* Reading, MA: Addison-Wesley.

DeLong, David. 1988. *"Computers in the Corner Office."* The New York Times.

DeMarco, Tom. 1978. *Structured Analysis and System Specification*. Englewood Cliffs, NJ: Prentice-Hall.

DeRossi, Claude J. and David L. Hopper. 1984. *Software Interfacing; a User and Supplier Guide*. Englewood Cliffs, NJ: Prentice-Hall.

Dierkes, Meinolf and Ute Hoffman. 1992. *New Technology at the Outset: Social Forces in the Shaping of Technological Innovations*. Frankfurt/New York: Campus Verlag.

Dreyfus, Herbert. 1979. *What Computers Can't Do*. New York: Harper and Row.

Dubin, Robert. 1956. "Industrial Workers' Worlds: A Study of the 'Central Life Interests' of Industrial Workers." *Social Problems* 3(January): 131-142.

Edwards, Richard C. 1979. *Contested Terrain: The Transformation of the Workplace in the Twentieth Century*. New York: Basic Books.

Ehn, Pelle. 1989. *Work-Oriented Design of Computer Artefacts*. Stockholm: Arbetslivscentrum.

Ellul, Jacques. 1964. *The Technological Society*. New York: Vintage Books.

Ferguson, Eugene S. 1979. "The American-ness of American Technology." *Technology and Culture* 20(1): 3-24.

Fischer, Frank and Carmen Sirianni. 1984. *Critical Studies in Organisation and Bureaucracy*. Philadelphia: Temple University Press.

Flood, Ann B. and W. Richard Scott. 1987. "Professional Power and Professional Effectiveness: The Power of the Surgical Staff and the Quality of Care." *Hospital Structure and Performance*. Baltimore, MD: The Johns Hopkins University Press.

Florman, Samuel C. 1976. *Existential Pleasures of Engineering*. New York: St. Martin's Press.

Fombrun, C. J. 1986. "Structural Dynamics Within and Between Organisations." *Administrative Science Quarterly* 31:403-421.

Form, William. 1983. "Sociological Research and the American Working Class." *The Sociological Quarterly* 24:163-184.

Fox, Alan. 1980. "The Meaning of Work." *The Politics of Work and Occupations*, edited by Geoff Esland and Graeme Salaman. Toronto: University of Toronto Press.

Franke, Richard H. and James D. Kaul. 1978. "The Hawthorne Experiments: First Statistical Interpretation." *American Sociological Review* 43(October).

Franke, Richard H. 1979. "The Hawthorne Experiments: Review." *American Sociological Review* 44.

Franz, Charles R. and Daniel Robey. 1984. "An Investigation of User-led System Design: Rational and Political Perspective ." *Communications of the ACM* 27(12, December):1202-1209.

Freeman, Peter. 1985. Concepts for Understanding Design Methods. Unpublished paper. University of California, Irvine: Department of Information and Computer Science.

Friedman, Andrew L. 1989. *Computer Systems Development: History, Organisation and Implementation*. Toronto: John Wiley & Sons.

Giddens, Anthony. 1984. *The Constitution of Society*. California: University of California Press.

Gillespie, Richard. 1991. *Manufacturing Knowledge: A History of the Hawthorne Experiments*. New York: Cambridge University Press.

Gordon, David, Richard Edwards and Michael Reich. 1982. *Segmented Work, Divided Workers*. New York: Cambridge University Press.

Graham, Margaret and Stephen R. Rosenthal. 1986. "Flexible Manufacturing Systems Require Flexible People." *Human Systems Management*.

Grint, Keith and Steve Woolgar. 1992. "Computers, Guns, and Roses: What's Social About Being Shot?" *Science, Technology, & Human Values* 17(3, Summer): 366-380.

Grudin, Jonathan. 1991. "Systematic Sources of Suboptimal Interface Design in Large Product Development Organisations." *Human-Computer Interaction* 6:147-196.

Gruneberg, Michael and Toby Wall. 1984. *Social Psychology and Organisational Behaviour*. New York: John Wiley & Sons.

Hall, Richard H. 1987. *Organisations: Structures, Processes, & Outcomes*. Englewood Cliffs, NJ: Prentice-Hall Inc.

Hamilton, Richard F. 1972. *Class and Politics in the United States*. New York: John Wiley.

Hamper, Ben. 1992. Rivethead: *Tales From the Assembly Line*. New York: Warner Books.

Harris, Catherine L. 1985. "Information Power: How Companies Are Using New Technologies to Gain a Competitive Edge." *Business Week*, October, 108-114.

Harrison, Bennett. 1991. "The Failure of Worker Participation." *Technology Review* 94(1, January).

Hart, Ann Weaver. 1990. "Work Redesign: A Review of Literature for Education Reform." *Advances in Research and Theories of School Management and Educational Policy* 1:31-69.

Hauser, John and Don Clausing. 1988. "The House of Quality." *The Harvard Business Review*, May-June, 63-73.

Hedberg, Bo and Enid Mumford. 1975. "The Design of Computer Systems: Man's Vision of Man as an Integral Part of the System Design Process." *Human Choice and Computers*. New York: American Elsevier.

Hicks, James O. 1976. *Management Information Systems: A User Perspective*. New York: West Publishing Company.

Hirschheim, Rudy A. 1985. "User Experience with Assessment of Participative Systems Design." *MIS Quarterly*, December, 295-303.

Hirschheim, Rudy and Heinz K. Klein. 1989. "Four Paradigms of Information Systems Development." *Communications of the ACM* 32(10, October):1199-1215.

Hirschhorn, Larry. 1984. *Beyond Mechanisation: Work and Technology in a Postindustrial Age*. Cambridge, MA: MIT Press.

Hughes, Thomas P. 1987. "The Evolution of Large Technological Systems." In *The Social Construction of Technological Systems*, edited by Wiebe E. Bijker, Thomas P. Hughes and Trevor Pinch. Cambridge: The MIT Press.

Ives, Blake, Margaret H. Olsen and J. Baroudi. 1983. "The Measurement of User Information Satisfaction." *Communications of the ACM* 26:785-793.

Ives, Blake and Gerard Learmonth. 1984. "The Information System as a Competitive Weapon." *Communications of the ACM* 27:1193-1201.

Ives, Blake and Margaret H. Olsen. 1984. "User Involvement and MIS Success: A Review of Research." *Management Science* 5.

Janda, A. 1983. *Human Factors in Computing Systems.* New York: North Holland.

Johnson, Bonnie McDaniel and Ronald E. Rice. 1987. *Managing Organisational Innovation: The Evolution from Word Processing to Office Information Systems.* New York: Columbia University Press.

Jones, Stephen R. G. 1992. "Was There a Hawthorne Effect?" *American Journal of Sociology* 98(3, November):451-468.

Kaiser, K. M. and R. P. Bostrom. 1982. "Personality Characteristics of MIS Project Teams: An Empirical Study and Action-research Design." *MIS Quarterly* 6:43-60.

Keen, Jeffrey S. 1981. *Managing Systems Development.* New York: North Holland.

Kiechel, Walter III. 1992. "When Management Regresses." *Fortune,* 9 March, 157-158.

Kimberly, John R. and Michael J. Evanisko. 1981. "Organisational Innovation: The Influence of Individual, Organisational, and Contextual Factors on Hospital Adoption of Technological and Administrative Innovations." *Academy of Management Journal* 24(4): 689-713.

King, William R. and Jamie I. Rodriguez. 1981. "Participative Design of Strategic Decision Support Systems: An Empirical Assessment." *Management Science.*

Kling, Rob. 1980. "Social Analysis of Computing: Theoretical Perspectives in Recent Empirical Research." *Computing Surveys* 12:61-110.

Kling, Rob and Suzanne Iacono. 1984. "The Control of Information Systems After Implementation." *Communications of the ACM* 27:1218-1226.

Koppel, Ross, Eileen Appelbaum and Peter Albin. 1988. "Implications of Workplace Information Technology: Control, Organisation of Work and the Occupational Structure." *Sociology of Work* 4:125-152.

Kroninger, Stephen. 1992. "Deconstructing the Computer Industry." *Business Week,* November 23, 90-100.

Kumar, Kuldeep and Niels Bjorn-Andersen. 1990. "A Cross-Cultural Comparison of IS Designer Values." *Communications of the ACM* 33(5):528-538.

Leana, Carrie R., Edwin A. Locke and David M. Schweiger. 1990. "Fact and Fiction in Analysing Research on Participative Decision Making: A Critique of Cotton, Vollrath, Froggatt, Lengnick-Hall, and Jennings." *Academy of Management Review* 15(1):137-146.

Leibenstein, Harvey. 1987. *Inside the Firm: The Inefficiencies of Hierarchy.* Cambridge, MA: Harvard University Press.

Ling, Richard. 1988. "Alternative Valuation of Computer Software: A Case Study Examining the Software Production Process." Unpublished paper. Oslo, Norway: Gruppen for Ressursstudier, Resource Policy Group.

Livesay, Jeff. 1989. "Structuration Theory and the Unacknowledged Conditions of Action." *Theory, Culture & Society* 6: 263-92.

Locke, Edwin, David Schweiger and Gary P. Latham. 1979. "Participation in Decision Making: One More Look." *Research in Organisational Behaviour*, edited by B. Staw. Greenwich, CT: JAI Press.

Locke, Edwin A. and David M. Schweiger. 1979. *Participation in Decision-Making: One More Look*. Greewich, CT: JAI Press, Inc.

Locke, Edwin A. 1984. "Job Satisfaction." *Social Psychology and Organisational Behaviour*, edited by M. Gruneberg and T. Wall. Chichester: John Wiley & Sons Ltd.

Locke, Edwin A., David M. Schweiger and Gary P. Latham. 1987. "Participation in Decision Making: When Should It Be Used?" *Organisational Behaviour and the Practice of Management*. Glenview, IL: Scott, Foresman and Company.

Lowith, Karl. 1964. *From Hegel to Nietzsche: The Revolution in Nineteenth-century Thought*. New York: Holt, Rinehart and Winston.

Lucas, Henry C. Jr. 1985. *The Analysis Design and Implementation of Information Systems*. New York: McGraw Hill.

Lund, Robert T., Albert B. Bishop, Anne E. Newman and Harold Salzman. 1993. *Designed to Work: Production Systems and People*. Englewood Cliffs, NJ: Prentice-Hall.

Magjuka, Richard. 1990. "Participation in Decision Making: An Empirical Analysis." *Advances in Research and Theories of School Management and Educational Policy* 1:237-277.

Majchrzak, Ann and Harold Salzman. 1989. "Introduction to the Special Issue: Social and Organisational Dimensions of Computer-Aided Design." *IEEE Transactions on Engineering Management* 36(3, August): 174-179.

March, James G. and Herbert A. Simon. 1958. *Organisations*. Wiley and Sons.

Markus, M. Lynne and Jeffery Pfeffer. 1983. "Power and the Design and Implementation of Accounting and Control Systems." *Accounting, Organisations, and Society* 8:205-218.

Markus, M. Lynne. 1984. *Systems in Organisations: Bugs and Features*. Boston: Pitman Publishing Inc.

Marx, Karl. 1977. *Capital*. New York: Random House.

Meiksins, P. F. and J. M. Watson. 1989. "Professional Autonomy and Organisation Constraint: The Case of Engineers." *Sociological Quarterly*, December.

Meyer, Marshall W. and Associates. 1978. *Environments and Organisations*. San Francisco: Jossey-Bass Publishers.

Meyer, John W. and Brian Rowan. 1991. "Institutionalised Organisations: Formal Structure as Myth and Ceremony." In *The New Instituionalism in Organisational Analysis*, edited by Walter W. Powell and Paul J. DiMaggio. Chicago: The University of Chicago Press.

Miller, Delbert C. and William H. Form. 1980. *Industrial Sociology: Work in Organisational Life*. New York: Harper & Row Publishers, Inc.

Mirvis, Philip H. and Edward E. III Lawler. 1983. "Systems Are not Solutions: Issues in Creating Information Systems That Account for the Human Organisation." *Accounting, Organisation, and Society* 8:175-190.

Mumford, Lewis. 1934. *Technics and Civilization*. Harcourt, Brace & World, Inc.

Mumford, Enid. 1981. *Values, Technology and Work*. Boston: Nijhoff Publishers.

National Center on Education and the Economy. 1990. America's Choice: High Skills or Low Wages-the *Report of the Commission on the Skills of the American Workforce*. Rochester, NY: National Center on Education and the Economy.

Newman, Michael and Faith Noble. 1990. "User Involvement as an Interaction Process: A Case Study." *Information System Research* 1(March):89-113.

Noble, David F. 1977. *America by Design*. New York: Oxford University Press.

Oerup, Agneta. 1989. "Socio-technical Design of Computer-assisted Work: A Discussion of the ETHICS and Tavistock Approaches." *Scandinavian Journal of Information Systems* 1(August):43-71.

Orlikowski, Wanda J. 1992. "The Duality of Technology: Rethinking the Concept of Technology in Organisations." *Organisation Science* 3(3, August):398-427.

Ortmann, GŸnter, A. Windeler, A. Becker, H.-J.Schulz. 1990. *Computer und Macht in Organisationen: Mikropolitische Analysen*. Opladen: Westdeutscher Verlag.

Osterman, Paul. 1988. *Employment Futures: Reorganisation, Dislocation, and Public Policy*. New York: Oxford University Press.

Palmer, Bryan. 1975. "Class, Conception and Conflict: The Thrust for Efficiency, Managerial Views of Labor and the Working Class Rebellion, 1903-22." *Review of Radical Political Economics*.

Pava, Calvin. 1983. *Managing New Office Technology: An Organisational Strategy*. New York: The Free Press.

Perrow, Charles. 1965. "Hospitals: Technology, Structure, and Goals." *Handbook of Organisations*, edited by James G. March. Chicago: Rand McNalley.

Perrucci, Robert and Joel E. Gerstl. 1969. *Profession Without Community: Engineers in American Society*. New York: Random House.

Petroski, Henry. 1982. *To Engineer is Human: The Role of Failure in Successful Design*. New York: St. Martin's Press.

Pfeffer, Jeffrey. 1978. "The Micropolitics of Organisations." *Environments and Organisations*. San Francisco: Jossey-Bass Publishers.

Paint, Jean. 1967. *The Child's Conception of the World*. Totawa, NJ: Littlefield.

Powell, Walter W. and Paul J. DiMaggio. 1991. *The New Institutionalism in Organisational Analysis*. Chicago: Chicago Press.

Rammert, Werner. 1992. "Research on the Generation and Development of Technology: The State of the Art in Germany." pp. 62-89 in *New Technology at the Outset*, edited by Meinolf Dierkes and Ute Hoffmann. Frankfurt/New York: Campus Verlag.

Rauner, Felix, Lauge Rasmussen and J. M. Corbett. 1988. "The Social Shaping of Technology and Work: Human Centred CIM Systems." *AI & Society* 2:47-61.

Rauner, Felix and Klaus Ruth. 1990. "Perspectives of Research in 'Industrial Culture'." *Ergonomics of Hybrid Automated Systems II*.

Robey, Daniel and Dana Farrow. 1982. "User Involvement in Information System Development: A Conflict Model and Empirical Test." *Management Science* 28(1, January):73-85.

Robey, Daniel and M. Lynne Markus. 1984. "Rituals in Information System Design." *MIS Quarterly*, March, 5-15.

Rockart, John F. and David W. DeLong. 1982. "The CEO Goes On-line." *Harvard Business Review*, January/February, 82-88.

Rosenberg, Nathan. 1976. *Perspectives on Technology*. New York: Cambridge University Press.

Rosenthal, Stephen R. and Harold Salzman. 1990. "Hard Choices About Software: The Pitfalls of Procurement." *Sloan Management Review*, Summer.

Rosenthal, Stephen R. 1992. *Effective Product Design and Development*. Homewood, IL: Business One Irwin.

Russell, Raymond. 1988. "Forms and Extent of Employee Participation in the Contemporary United States." *Work and Occupations* 15(4, November):374-395.

Salzman, Harold. 1989. "Computer-aided Design: Limitations in Automating Design and Drafting." *IEEE Transactions on Engineering Management* 36(4).

Salzman, Harold and Robert Lund. 1995. "Engineering principles and production technology: automation and skill-based design in the U.S," *International Journal of Vehicle Design* v.16, No4/5

Sashkin, Marshall. 1984. "Participative Management is an Ethical Imperative." *Organisational Dynamics*, Spring, 5-22.

Schniederman, Ben. 1987. "Designing Computer Systems for People." *Contemporary Psychology* 32:779-780.

Schweiger, David and Carrie Leana. 1986. "Participation in Decision Making." *Generalising from Laboratory to Field Settings*, edited by Edwin Locke. New York: Lexington Books.

Scott, W. Richard. 1987. *Organisations Rational, Natural, and Open Systems*. Englewood Cliffs, NJ: Prentice-Hall.

Segal, Howard P. 1985. *Technological Utopianism in American Culture*. Chicago: The University of Chicago Press.

Sewell, William H. Jr. 1992. "A Theory of Structure: Duality, Agency, and Transformation." *American Journal of Sociology* 98(1, July):1-29.

Stinchcombe, Arthur L. 1990. *Information and Organisations*. Berkeley: University of California Press.

Strassman, W. Paul. 1959. "Creative Destruction and Partial Obsolescence in American Economic Development." *The Journal of Economic History* 19(3, September).

Sykes, A. J. M. 1965. "Economic Interest and the Hawthorne Researches." *Human Relations* 18.

Tait, Peter and Iris Vessey. 1988. "The Effect of User Involvement on System Success: A Contingency Approach." *MIS Quarterly*, March, 91-108.

Taylor, Fredrick W. 1911. The Principles of Scientific Management. New York: W. W. Norton. Texas Center for Production and Quality of Work Life. 1983. An Assessment of US Work Improvement Case Studies and Lessons: Learnings from Other Organisations Change Efforts. unpublished paper. University of Texas.

Thomas, Robert J. 1993. *What Machines Can't Do: Politics and Technology in the Industrial Enterprise*. Berkeley: University of California Press.

Thuesen, G. J. and W. J. Fabrycky. 1989. *Engineering Economy*. Prentice-Hall.

Trist, Eric L. 1981. "The Evolution of Socio-technical Systems." *Ontario Quality of Working Life Center.*

Trist, Eric and Hugh Murray. 1990. "Historical Overview; the Foundation and Development of the Tavistock Institute." The Social Engagement of Social Science: *A Tavistock Anthology,* edited by Eric Trist and Hugh Murray. Philadelphia: The University of Pennsylvania Press.

Turkle, Sherry. 1984. *The Second Self: Computers and the Human Spirit.* New York: Simon and Schuster.

Turner, Jon A. 1981. "Achieving Consensus on Systems Requirements." *Systems, Objectives, Solutions.*

Walton, Richard E. 1985. "From Control to Commitment in the Workplace." *Harvard Business Review,* March-April, 77-84.

Wells, Donald M. 1987. *Empty Promises: Quality of Working Life Program and the Labor Movement.* New York: Monthly Review Press.

Weltz, Friedrich. 1991. "Der Traum Von Der Absoluten Ordnung Und Die Doppelte" *Wirklichkeit Der Unternehmen* (The Ideal of Absolute Order and the Dual Reality of the Firm)." Betriebliche Sozialverfassung Und Veraenderungsdruk. Berlin.

Whalley, Peter. 1986. *The Social Production of Technical Work: The Case of British Engineers.* State University of New York Press.

Winner, Langdon. 1977. *Autonomous Technology: Technics-out-of-control as a Theme in Political Thought.* Chicago: The University of Chicago Press.

Winograd, Terry and Fernando Flores. 1986. *Understanding Computers and Cognition: A New Foundation for Design.* Reading, MA: Addison-Wesley Publishing Company, Inc.

Woolgar, Steve. 1991. "The Turn to Technology in Social Studies of Science." *Science, Technology, & Human Values* 16(1, Winter):20-50.

Zuboff, Shoshana. 1988. *In the Age of the Smart Machine: The Future of Work and Power.* New York: Basic Books.

Zussman, Robert. 1985. *Mechanics of the Middle Class: Work and Politics Among American Engineers.* Berkeley: University of California Press.

6 Computer Networking and Interdisciplinary Research

MARGIT POHL
VIENNA TECHNICAL UNIVERSITY
VIENNA; AUSTRIA

Abstract:

The introduction of computer technology, especially computer networks, leads to fundamental economic and organisational change in modern enterprises. Compared to traditional technology the benefits of modern technology are indirect and difficult to quantify. Evaluation of investments should, therefore, not be restricted to cost/benefit analysis. An Organisation becomes more flexible and less hierarchical, and autonomous workgroups are developed. For a detailed analysis of these processes, it is necessary to apply a holistic, interdisciplinary approach as computer technology affects many aspects of modern working life.

Introduction

It is quite obvious that the introduction of computer technology leads to fundamental economic and organisational changes. But it is difficult to assess these changes because of the integrated and complex character of this technology. Traditional cost/benefit analysis assumes that costs and benefits originate in the same department. This assumption does not hold in the case of strategic information systems. It is perfectly possible that cost incurred during operations scheduling can lead to a better utilisation of machines or more generally to a greater flexibility on the shop floor. On the other, hand not all the effects of the introduction of network systems are positive. The introduction of network systems might devalue the qualification of middle management on the shop floor. There are tendencies that the autonomy on the shop floor is being restricted because of the constraints inherent in the system. To get a comprehensive picture of the changes brought about by information and communication technologies it is necessary to adopt an integral view. Therefore, the introduction of information and communication technology is often called "Systemic Rationalisation" (systemische Rationalisierung), whereas conventional technological change can be described as rationalisation that affects only a small part of the whole system. This latter kind of rationalisation is typical of the early phases of the development of computer technology. The creation of centralised information systems departments in the sixties did not influence the traditional organisational structure of companies. They just offered additional services. The realisation of CIM concepts on the other hand implies fundamental organisational and structural change. These concepts are not restricted to the integration of one company but also include means to achieve remote data processing (electronic data interchange). In Austria, local area networks are fairly common whereas electronic data interchange meets with some resistance.

There are a considerable amount of publications concerning design, analysis and implementation of computer systems. But the economic and organisational impact of these decisions has not been investigated very thoroughly. Common sense is very often the best strategy to decide on large investment projects. Representatives of large Austrian companies told us in interviews that they still use conventional methods of profitability assessment, even if they know that the systems have many effects that are not quantifiable. Most of them do not use any formalised methods to evaluate qualitative benefits. Usually, they only rely on expert opinion (OFNER & POHL).

A holistic, interdisciplinary approach to technology assessment implies that there are more fundamental difficulties with introducing strategic information systems than the problem with accounting. Companies usually are embedded in a certain competitive environment. This environment may change dramatically during the process of implementing a new information system, especially in dynamic, rapidly changing industries. And, in the case of large companies, the strategic investment itself may alter the business environment and thereby create uncertainty. There are several reasons for this phenomenon. The size of investments is considerably higher than for traditional technology, the lead time is longer and the impact on the organisation is more fundamental. There are at least two assumptions that can be made concerning the introduction of strategic information systems. First, evaluation of investments based on potential competitive impact gets more and more important compared to evaluation of investments based on cost. Second, the introduction of strategic information systems may lead to a more thorough organisational change than traditional technology. Network systems not only affect management structures and division of labour, but also the company's relation to suppliers and customers.

It is rather obvious that organisational changes have to take place but the empirical evidence on how these changes should look are contradictory. Several approaches to solve these problems can be distinguished. There are technology centred approaches which assume (at least implicitly) that organisations should adapt to existing technical solutions. These methodologies have led to numerous practical problems, especially with the employees involved. The concept of human-centredness, which assumes that any information system consists primarily of human beings, is supposed to overcome these problems. On a methodological level these two approaches are not so different as it may seem. And there are certain developments in business systems that cannot easily be interpreted in terms of the one theory or the other. In the sixties and early seventies computer systems in business environments were very centralised, but the work of the information systems departments did not influence other departments very much. Since the mid-seventies we can observe a combined tendency of centralisation and decentralisation. On the one hand, network systems permit information system departments to control the work of other departments to a large extent, on the other hand the same systems allow users to get access to more information than ever before. The extent of centralisation and decentralisation depends on the tradition of companies and on processes of negotiation between different groups in the company rather than on technological constraints. It is, for example, part of the tradition of a company if programming outside of the information systems department is allowed or not. But it is not clear if decentralisation as such is human-centred or not.

Overview

In the following I want to describe some important research areas in social sciences, which are relevant for the overall evaluation of information systems. The selection of these topics is, to some extent, arbitrary, because there is no comprehensive theory of economic and social impacts of information technology. This is not surprising as this topic requires an interdisciplinary approach which includes many different disciplines as for example computer science, applied economics, economics, sociology or psychology. Basically, there is no argument against the co-operation between these disciplines but the realisation of this project seems rather complicated because of theoretical and methodological differences. Most of the empirical research in this area is restricted to one of these disciplines and addresses only part of the problem. It is therefore not possible to give a comprehensive description of a theory or optimal solutions for the problems mentioned above.

The following three points are just a few of the possible relevant topics. But I have tried to integrate economic and sociological approaches to avoid a restriction to problems of accounting and to point out the possibility of alternative solutions.

- Assessment of Profitability
- The Question of Competitive Advantages
- Organisation and Co-ordination.

It is very difficult to assess the profitability of information and communication technologies by traditional techniques based on cost. This is due to the integral character of these new technologies which results in indirect and unquantifiable benefits. There is a considerable consensus on this question. Moor (1989), for example, states that the complexity, the strategic importance, the extended period of planning and the unquantifiable effects of computer technology imply that other instruments than cost/benefit analysis should be used in addition to evaluate investment decisions. In his view, the necessity to quantify all the data used is the most debatable point about cost/benefit analysis.

It is possible to distinguish between four different kinds of benefits of network systems:

Direct, Quantifiable Benefit

Direct benefit is a consequence of the exchange of information as such. In this context it is not relevant if the quality of information is improved. Generally, direct benefits lead to a reduction of cost through staff reduction and saving of time. To avoid entering the same data in different departments repeatedly is a good example for this category. As staff reduction is an important element of this kind of benefit it is obvious that it is easily quantifiable.

Direct Benefit, Difficult to Quantify

This effect is again independent of the quality of information but its monetary evaluation is very difficult. An estimation of the reduction of errors in information exchange for example is quite difficult to achieve as it is hard to ascertain how many errors there would have been without a network system. In addition, the character and volume of tasks usually changes with the introduction of information and communication technology which makes any comparison even more doubtful.

Indirect,Quantifiable Benefit

Indirect benefits originate from a better quality of the information available for decision making. The reduction of the machining time as a consequence of the introduction of network systems would be an example for this kind of benefit. This reduction can be achieved because precise information on capacity utilisation or operations scheduling can be accessed via the information system at any time. It is an increase in quality because similar information would not have been available before the introduction of an information and communication system. Today, urgent orders can be executed very quickly and capacity utilisation can be increased because of the greater flexibility of business systems. The reduction of the machining time as such can be quantified quite easily. The problems arise when you start to compare the state before and after the introduction of an information system. This variable can only be interpreted in a meaningful way when the scope of products remains the same throughout the process which is usually not the case. In addition, it might not be possible to identify a causal relationship between the reduction of the machining time and the introduction of a network system because this reduction might also be the effect of organisational change related to the introduction of the network system. In other words, is it really the improved quality of information which is responsible for the benefit and how can I measure this improvement? And if this relationship exists, is it linear or not? It is obvious that even if the machining time is easily quantifiable it is only an estimate for the benefit of a new information system.

Indirect Benefit, Difficult to Quantify

Estimates for this category are the most dubious because the variables are not only difficult to quantify but they also get inaccurate because of the problem of assessment of information quality. A product design which is adapted to the requirements of the production process can be attributed to the introduction of an information and communication system only with difficulties. Furthermore, this variable is difficult to quantify. It is probably possible to determine whether the design is more adapted to the requirements than before but it is certainly not possible to identify the amount of improvement.

Horvath (1989) also distinguishes between internal and external effects. Internal effects refer to the company itself, external effects to its competitive environment. External effects relate to variables like flexibility or quality of the products. Horvath developed a model of benefits which is supposed to make the process of benefit assessment more precise and meaningful:

Effects of network systems

Direct Benefits

Easily quantifiable, internal effects:
- Avoidance of repeated entry of the same data
- Reduction of delay
- Reduction of expense for modifications
- Reduction of exchange of documents
- Reduction of demand for memory/storage
- Reduction of expense for traditional information exchange.

Difficult to quantify, external effects:
- Reduction of interruptions of the work flow
- Reduction of routine tasks
- Reduction of expense for prototyping
- Increase of standardisation
- Reduction of errors during information exchange
- Increase of system security
- Improvement of conditions for the introduction of a three-shift-system
- Clear responsibilities for certain tasks.

Indirect Benefits

Easily quantifiable, internal effects:
- Higher capacity utilisation
- Reduction of machining time
- Reduction of stocks
- Reduction of workshop supply
- Higher quality of products.

Difficult to quantify, internal effects:
- Product design adapted to requirements of the production process
- Consideration of logistic aspects during the planning process
- Reduction of expense for modifications

- Reduction of time for the introduction of new products
- Increased consciousness for the goals of the company as a whole
- Reduction of expense for prototyping
- Higher acceptance of the usage of information.

Difficult to quantify, external effects:
- Higher flexibility
- Higher quality
- Reduction of lead time.

(*Source*: Horvath 1989, p.140)

One possibility to include qualitative criteria into the assessment of investments are qualitative methods like the utility value analysis. This method also uses subjective preferences as a basis for the decision process. The utility value analysis consists of three steps:
- Preparation of a hierarchy of the company's goals
- Weighting the goals
- Assessment of different alternatives according to the degree of fulfilment of the goals

The results of this assessment are supposed to be added to traditional methods of accounting in order to avoid the acquisition of a system that is cheaper but produces less (qualitative) benefit than a more expensive one.

The application of the utility value analysis is a controversial question. Several scientists state that the subjectivity of evaluation leads to inefficient decisions:

Several scientists state that the subjectivity of evaluation leads to inefficient decisions. Utility value analysis is supposed to overrate qualitative dimensions which leads to an illusory impression of cost effectiveness. In some cases its application makes some sense but it would be impossible to evaluate the overall cost effectiveness of CIM investments. - (Horvath 1989).

Horvath is right insofar as every technological innovation must lead to a reduction of cost. Flexibility or quality of products as such are no primary goals for companies. On the contrary, higher flexibility is considered a burden rather than an advantage (Ofner & Pohl, 1991). Most companies do not utilise the potential flexibility of their information and communication systems. One of the reasons mentioned by the managers of Austrian companies was that the administrative overhead costs are so high that a large scope of products cannot be produced effectively. On the other hand, Horvath's preference for quantitative variables may result in a fictitious accuracy that has no basis in reality.

> Developing a strategic application - intended to make a company more flexible, more responsive to customer needs, or more able to adapt to rapidly changing conditions in the competitive environment - is fundamentally different from investments undertaken to automate the back office to reduce expenses or increase capacity. Alternative techniques for evaluating the business case for strategic systems have been developed and have worked well in practice. (Clemons 1991, p.23).

Clemons states that the evaluation of strategic investments should not be restricted to traditional cost/benefit analysis but should also include considerations on competitive advantages because strategic investments (e.g. the travel agent reservation systems of major airlines) usually have a fundamental influence on the market position of companies. He points out that there are complex methods of algorithm and data structure design, there are tools for the management of software development and statistical quality control, and there is much improvement in user interface design. But there are scarcely any formalised methods or guidelines to support senior managers when they decide on strategic information systems.

Many decisions seem fairly obvious from today's perspective even if the process to reach this decision was by no means easy. An example for this might be American Airline's travel agent reservation system "Sabre". The company started to develop its own system in a situation when their main competitor United Airlines had already introduced his own system to the market. By adopting an aggressive marketing strategy American Airlines succeeded to get a major market share. The market value of "Sabre" is now higher than that of American Airlines' core business. On the other hand, many similar decisions were not successful.

It is obvious that in a situation like that there is an urgent need for methods to support senior manager's decisions on strategic information systems. But the available techniques have not kept pace with the development of information technology. There is still a tendency rather to make investment decisions based on precision than on accuracy. In other words, there is a preference for exact numbers at the expense of a realistic description of actual conditions. This preference for exact numbers is the more dubious as strategic information systems are very often introduced in situations of great uncertainty. Strategic information systems usually have a long lead time. As a consequence it is often impossible to predict market conditions for the time the system is supposed to be realised. This applies even more to dynamic, deregulated industries. Frequently, it is not possible to get objective and exact data to support an investment decision.

Exact numbers might even give you a distorted picture of reality. Most decisions for strategic information systems include a great deal of risk because of the uncertain elements of the business environment. To make a secure decision based on exact predictions of potential costs and benefits under such circumstances very often means to select a conservative solution based on the notion that the status quo can be maintained. In the long run this strategy can lead to a diminishing market share.

The goal of any guideline to support senior management in their strategic investment decisions should be to avoid the tendency to conservative decisions and at the same time to minimise the risk. In his article Clemons describes several principles which can be used for decision support. These principles are a result of empirical research with major corporations, mainly in the USA.

Sometimes, it may be possible to make a rational and analytical decision even if it is not possible to use conventional methods of accounting. The experience of social sciences shows that rank scales can be used successfully to determine the quality of several solutions. This method can also be applied to support the selection process between alternative investment projects.

In some cases exact numbers can be computed, but it is impossible to decide which of a number of scenarios is going to happen. In these cases it is advisable to use a sensitivity analysis, a method used for dealing with uncertainty in decision problems. Several different possible scenarios are computed under the assumption that the decision maker introduced his/her preferred system. If the output variables do not vary too much for the different scenarios the system is recommended, otherwise additional investigation is necessary.

Decisions on strategic information systems can include different kinds of risks. The alternatives related to the risks have to be weighed carefully. Usually, there is a conflict between secure, conservative decisions and risky decisions. In the first case, the company would select a cheap system that can be realised without any problems. This selection is risky in the sense that it may lead to a loss of market shares because the system specifications are based on the current situation. As a consequence the system will probably be outdated when it is introduced. In the second case, an expensive system that might never be realised in its full functionality will be acquired. However, this alternative takes into consideration dynamic market structures and changing demand.

One important element of strategic decision making processes is the motivation of the employees affected by the new system. It is necessary to discuss possible alternatives with these employees and consider critique in advance. In this context it is easier to convey the advisability of a strategic investment decision even if no conventional assessment of profitability has been made. A crucial point for this strategy is the credibility of the project leader. One possible method to minimise risk even here would be to divide the project into several modules.

Strategic investment decisions have to be based on the notion that no new technology can guarantee a competitive advantage indefinitely. Every technology can also be applied by the competitor. The only way to stay successful is the creative integration of new and efficient technology and nontechnological assets, as for example scope of services offered to the customers.

If a firm is not provided with such nontechnological assets it is advisable to follow a co-operative strategy. This applies all the more if the competitor dominates the market because of his size. As the costs for system's development are very high, companies can reduce them through co-operation. They can achieve necessary economies of scale and get access to important resources. Because of its positive economic effects this kind of co-operation becomes more and more common.

An important variable that has to be considered when deciding on strategic information systems is the changing status quo. Secure profits in the present often lead to a postponement of investments necessary for the future. This strategy is based on the implicit assumption that the current status quo will be maintained indefinitely. This assumption does not hold for dynamic markets. Even if a company decides against an advanced technology a competitor might successfully introduce it. Clemons calls this "the trap of the vanishing status quo". As a consequence, discussions on strategic investments should always take into account potential losses. The most important criterion for this strategy are stable market shares rather than high profits.

Clemon's discussion of necessary changes in deciding on strategic information systems has shown that organisational or sociological aspects get more and more important even for such a seemingly remote area as accounting. Motivation of the employees, tradition of firms or co-operative strategies play an important role for strategic investment decisions. Therefore, every efficient introduction of an information system has to be accompanied by some kind of organisational analysis.

A traditional definition of a computer would describe it as a programmable, electronic machine that processes and stores information. This definition is not valid any more as computer systems changed fundamentally during the last years. An important aspect of modern computer systems is the integration of single computers - the network system. Network systems permit efficient decentralised access to large amounts of information. During the seventies the computer could be aptly described by the "number cruncher" metaphor. Today, the computer effectuates important co-ordination activities between different departments in companies. Therefore, the concepts of co-ordination and networking cannot be discussed independently. Another effect of network systems is a change in the division of labour. Clerks in insurance companies for example have to perform a much broader scope of tasks after the introduction of network systems than before. In former times clerks specialised in certain types of insurances. Nowadays they are able to advise a customer on any kind of insurance because of efficient computer support.

The introduction of computer systems might lead to a decrease in competence for the middle management because the task of co-ordination of workforce or scheduling of operations can be performed by the network system. In the USA this results in comparatively flat hierarchies. In Austria this phenomenon cannot be observed (Dell'Mour et al 1991). But there might be tendencies pointing in this direction.

Malone and Rockart (1991) called this phenomenon a first-order effect because computer technology directly affects conditions of work. Second-order effects are related to the concept of co-ordination. Co-ordination gets cheaper because of the use of network systems. This leads to staff reduction and to the development of completely new services (e.g. the airline's travel agent reservation systems). Third-order effects describe company structures characterised by intensive co-ordination. Just-in-time production in the automobile industry is an example for this kind of intensive co-ordination. Co-ordination in this case is not restricted to a single firm but integrates several companies.

Information technology stimulates highly co-operative systems. As a consequence, organisational structures in companies change. Economic units are supposed to decrease in size. Companies tend to reduce their scope of production activities and rather buy goods and services on the market. Many decisions are therefore not made in the firms but on the market. Malone and Rockart consider this phenomenon as a characteristic of current decentralisation processes.

Generally, the introduction of information technology has a fundamental impact on company structures. Strategic investment decisions under conditions of uncertainty raise a whole lot of new questions. The problem of retaining or increasing market shares is an important criterion for these decisions. The introduction of highly co-ordinated structures in and between firms results in new supply and demand policies and alternative management concepts.

Assessment of profitability

Methods

Biethahn et al (1990) developed a concept of integral information management. An important part of their concept is an assessment of current methodologies used to evaluate investment decisions on information systems:

They distinguish between different methodologies:
One-dimensional techniques
- static techniques
- dynamic techniques
- Multi-dimensional techniques

Static techniques have severe disadvantages when they are used to evaluate investments in information systems. Frequently, it is not possible to determine the benefits for single investment objects. Besides, static techniques do not consider benefit fluctuations. Dynamic methods were devised to overcome this disadvantage. Biethahn et al criticise dynamic methods as follows:

"Dynamic methods were devised to overcome this disadvantage but Biethahn et al point out that dynamic methods have their shortcomings as well. The most fundamental one is that such models assume that it is possible to have perfect foresight. It is obvious that this is not the case." (Biethahn et al 1990, pp.234-235).

Utility value analysis

The utility value analysis is an alternative technique to evaluate the usability of information and communication systems. The decision for a complex information system depends on many different aspects, some of which cannot be evaluated in monetary terms. The utility value analysis consists of an evaluation of all these aspects. The weighted sum of these variables is then used as a guideline to select between different alternatives.

One of the characteristics of the utility value analysis is the use of subjective judgements of decision makers. This feature has led to a very controversial discussion. While Horvath (1989) criticises the application of the utility value analysis, Zolles (1979) argues for a careful utilisation. Zolles justifies his view by pointing out that the utility value analysis when applied appropriately makes subjective evaluations transparent. These same subjective evaluations also influence traditional numerical methods but in the latter case they cannot be detected. They are therefore not considered in the interpretation of the results.

The utility value analysis can be distinguished from the optimisation models of Operations Research insofar as it does not yield an optimal solution in the mathematical sense of the word as a result because subjective evaluations are included. But none the less in a certain sense it is possible to say that the utility value analysis maximises the subjective benefit of the decision maker that is expressed in his/her preference structure. Besides, the line between objective and subjective optimum is blurred because even in selection techniques based on mathematical models subjective information plays a certain role. Therefore, the utility value analysis is an appropriate method to evaluate complex information systems.

The utility value analysis can be roughly described as a process where the preference structure of the decision maker is determined and then used as a basis for the selection of the alternative with the highest values in target variables (Zangemeister 1976, p.45).

The utility value analysis is based on decision theory. One of the preconditions of this technique is the existence of a space consisting of a finite number of alternatives. The alternative with the highest benefits is selected. Decision theory assumes that all alternatives and their consequences are known at the time of the decision. Accordingly, the Utility value analysis consists of the following steps:

- Determination of the relevant goals
- Description of the relevant benefits
- Weighting of the alternatives
- Determination of partial benefits
- Computation of the overall benefit
- Verification of the result

Determination of the Relevant Goals

In the first phase the requirements are determined that an investment project should fulfil. These requirements must be specified exhaustively in order to make a verification of their fulfilment possible. The goals are criteria which are used to evaluate the alternatives. These criteria can have a varying degree of rigidity. Some of the criteria are an absolute must, others are only optional.

- minimum requirements
- desirable requirements
- unimportant requirements.

The minimum requirements must be met by every investment project at all events. If they are not fulfilled the project is at once rejected. Problems may arise when different conflicting goals must be fulfilled at the same time. Two conflicting goals for software systems would be user friendliness and a broad scope of functionalities. The more tasks a software system can fulfil the more difficult it is to learn. This shows that goals are not independent of each other. The hierarchy of requirements should be structured in a way to avoid such conflicts. It is usually not easy to guarantee this independence of goals.

Description of the Relevant Benefits

In this phase the alternatives are evaluated according to the degree of fulfilment of the criteria. This phase is often called impact analysis. The results of this analysis are usually represented as a matrix. One of its dimensions shows the different alternatives, the other the different criteria. The results can be represented on various scale levels. They can even be described verbally. The single criteria should be grouped according to meaningful concepts. These concepts should be developed in such a way that every area is taken into account according to its relative importance. It would be useless to develop a hierarchy with twenty criteria for the working memory and only five for the quality of the user interface design if the latter is supposed to be the most important feature of the system.

Weighting of the Alternatives

Weights for the criteria express the relative importance of the goals. If the criteria are already organised in groups this task is much easier. The criteria are weighted gradually in a top down procedure. It is advisable to ask more than one person to determine the weights as individual preferences can vary to a great extent.

Determination of Partial Benefits

For practical applications even the exact values are transformed to a certain number of points. In this way "objective" numbers are represented as "subjective" values. These points are multiplied with the corresponding weights. For every goal there must be a rule to decide on the degree of fulfilment. For example, it must be clear from the beginning if the number "5" means that the requirement is met or not.

Computation of the Overall Benefit

During this phase the partial benefits of the single alternatives are added up to get an overall result. This phase is also called value synthesis. The ranking of the different alternatives is based on the overall benefit. The alternative with the highest overall score should be preferred.

Verification of the Result

The result of the utility value analysis contains a lot of inaccurate information. Therefore, it is advisable to perform a plausibility check. A sensitivity analysis can find out how robust the results are if the initial values are changed. If the variations are high the result does not appear dependable.

On the one hand the utility value analysis is a feasible alternative to traditional techniques of accounting. On the other hand this method is certainly not an optimal solution. Biethahn et al (1990, pp.240-241) describe the problems as follows:

> On the other hand this method is certainly not an optimal solution even if there are certain advantages. Several components of utility value analysis can be influenced by subjective decisions: the choice of criteria or the weights assigned to individual criteria, for example. There is no possibility to derive any statements about the cost effectiveness of an investment project from utility value analysis. On the whole, subjectivity is considered an advantage because of the importance of motivational factors of the users of modern information systems.

Integration of markets

One of the most important features of modern information and communication technology is its ability to integrate different systems. This networking has a fundamental influence on relationships on national and international markets. These relationships increased considerably during the last years. Rada (1986) states that one of the main effects of modern communication technology is the transportability of services. This widespread availability of services is one of the most important factors of the increasing interdependence on markets. Some of the consequences of this development could be:

- increasing productivity in the service sector
- increasing transparency of markets
- changes concerning entrance barriers and internationalisation of the service sector.

The following table, put together by Rada, shows some of the effects of the introduction of information and communication technology which stimulate integration processes on national and international markets (the table is not quoted completely):

Shipping

Partial substitution of transport of goods by services (Substitution of shipped printed matter by sending it through telecommunications)

Other Transport (Air, Rail, Road, Inland, Waterways)

As above (Transport of mail is partly substituted by electronic mail).

Insurance and Reinsurance

More transportability (Customers will be able to order directly specific insurance through retailing points or terminals as is now the case in many airports, which offer flight insurance).

Banking and other Financial Services

More transportability (Automatic teller machines can reach many different places while providing 24-hour service. Telebanking is done now at corporate level and in some households. SWIFT will evolve into a worldwide trading network. Banks are also increasingly becoming information suppliers in finance, trade, and investment).

Wholesale and Retailing

Increased transparency of markets. Goods and services supermarkets (Easy access to price information and services could increase competition in this sector. Complex delivery systems are likely to be developed. In large supermarket chains, the trend is to offer also a "supermarket of services." Teleshopping is already used for a limited number of goods. This will evolve into a sort of "international mail order" catalogue with direct relationship between the consumer and main wholesale or retailer).

Construction Engineering (Management, Consulting, Design/Architecture)

More and highly transportable (The use of Computer Aided Design (CAD) systems and remote entry for calculations in centralised systems will increase transportability) (Rada 1986, pp64-66).

It is obvious that these tendencies affect very different parts of firms even if we assume that some of them will not be realised the way Rada describes. Krommenacker (1989) apparently has such a complex development in mind when he argues that a holistic view of this process must be adopted. His main point is that information technology has fundamental effects on the competitive position of companies. Three different types of these effects can be observed:

The development of information technology changes industrial structures. New companies develop which are vertically desegregated and rely heavily on their suppliers in important areas (sales, marketing, parts of the production process). This might be the organisational model of the enterprise of the future.

Information technology can be an important element of a strategy to realise competitive advantages. These competitive advantages can be achieved by reducing cost, by increasing the scope of products and by increasing flexibility. One method to reach this aim would be the integration of functions which were separated so far. An example for this development could be the fact that the activity of banks, insurance companies, and brokers cannot be distinguished easily any more. This process is stimulated by the information systems used in these areas.

Information technology creates completely new industries. The Credit Commercial de France, a French bank, introduced an electronic system that enables customers to raise a loan via a terminal which is installed at their flats.

The three trends described by Krommenacker still have to be corroborated by empirical data. Some of the assumed effects of information technology seem to be the outcome of wishful thinking rather than of scientific research. I have already mentioned the fact that there is a contradiction between cost reduction and increase of the scope of products or variants. Competitive advantages apparently can only be achieved either the one way or the other. This experience should be considered when potential impacts of information technology are discussed. Furthermore, it is certainly important not to forget Clemon's (1991) remark that information technology as such does not guarantee competitive advantage in the long run. It has to be combined with nontechnological assets. But it is important to point out that Krommenacker describes the relationship between cost, competitive position, organisation and new technology in a very plausible way.

Information technology and competitive position - a case study

One of the best known strategic information systems is "Economost" developed by McKesson Drug. Clemons (1991) quotes this system as an example for the application of sensitivity analysis. Johnston & Lawrence (1988) describe this system as a successful realisation of a "Value-Adding Partnership". They define this concept as follows:

> The term value-added chain comes from the field of microeconomics, where it is used to describe the various steps a good or service goes through from raw material to final consumption. Economics has traditionally conceived of transactions between steps in the chain as being arm's-length relationships or hierarchies of common ownership. Value-adding partnerships are an alternative to those two types of relationships. Usually, the partnerships first develop between organisations that perform adjacent steps in the chain. - (Johnston & Lawrence 1988, p.96).

A value-adding partnership is defined as the co-operation of several firms which represent different stages in the work process necessary to produce a certain good or service. The aim of this kind of co-operation is to improve the competitive position of the participating firms which are usually not very large. Sometimes there are several companies that represent one stage of the work-process. Thus, the structure can increase its flexibility because those companies that are over-utilised are supported by other members of the value-adding partnership. Another source of flexibility is the fact that there is no hierarchy and large administration. In this way the companies can adapt to the market very quickly without a large amount of bureaucratic effort. Ideally, bureaucracy should be replaced by creativity. Modern information technology is not absolutely necessary for this kind of organisation but it certainly is an advantage. This can be exemplified by the development of McKesson Drug.

The decision to introduce Economost was made in a situation of radical change in the drug wholesaling branch. The number of drug wholesalers decreased rapidly. The companies that disappeared from the market were mainly the ones which could not offer an electronic order system to their customers. McKesson Drug decided to develop such a system even if the outcome of this project was by no means certain. The company was successful in that the cost of processing the orders could be reduced considerably. This reduction could be achieved because different processes were accelerated. The processes of taking inventories in the single supermarkets, of entering orders, of answering orders and of transport are much quicker than before. Clemons (1991, p.29) describes the procedure as follows:

> It (Economost, M.P.) permits a drugstore employee to order by walking through the store with a barcode reader and recorder, and waving a wand at any item that appears to be in short supply. Goods arrive the next day, in the store's order quantity, laid out in accordance with the store's floor plan and complete with the store's prices. Numerous management reports are available, some without additional charge.

This description shows that Economost does not only increase the efficiency of McKesson Drug's administration but also that of the pharmacies and drugstores co-operating with the wholesaler. The managers of McKesson Drugs realised very early that an economic failure of their partners would lead to negative effects for their own company. Therefore, they tried to develop a system that would be profitable for the whole value-adding partnership.

The decision to develop Economost seems obvious now but was very difficult at first. The first prototype was installed 1973 at three sites. The reaction to this experiment was mixed. McKesson's salespeople felt massively threatened by the system. Part of the management argued that the system would lead to major price concessions. The only way to convince managers and users of the potential advantages of the system was the development of a mathematical model of the system to show its predicted behaviour in detail and to point out to managers that the price reductions could be more than compensated by cost reduction and increases in orders.

An important element of the model was the assumption that the customers would order considerably more because of the introduction of the electronic order entry system. This assumption could not be estimated accurately. Therefore, a sensitivity analysis was applied. The target variable of the system was return on investment (ROI). Critical variables that were changed for different scenarios were price level and sales effort. This model was computed for a number of assumptions. Economost still seemed attractive under most of the assumptions and was, therefore, implemented.

Clemons (1991, p.29) draws the following conclusion from this case study:

> Where there is uncertainty, but the possibility of quantifying some of the critical variables exists, financial analysis can be quite valuable. The sensitivity analysis that McKesson performed was essential in allowing them to reach their decision.

The timing both of the investment, and the reduction of uncertainty, can be critical. Here, McKesson was able to obtain information, refine its model, and rework the numbers before incurring full development costs. And had the initial response been very different from the predictions, the firm would have been able to stop the project with very little lost investments.

This case study shows different new tendencies in the evaluation process of investments in information technology. A creative application of mathematical modelling can provide rational decision support even if conventional methods of accounting do not succeed. Assumptions on the behaviour of the participants of the process had to be included in the model from the beginning. In this the model differs considerably from conventional accounting. Generally, the aspect of co-operation seems to be stimulated by the information system even if Johnston & Lawrence's description of the advantages of a value-adding partnership seems a little bit exaggerated. It remains to be seen if their model is sufficiently general to be applied to other companies.

Organisation

The McKesson Drug case study shows that new technology and changing conditions on the markets may bring about structural change in companies. Malone et al (1987) and Malone & Rockart (1991) state that new technologies cause firms to buy parts and services on the market rather than to produce them themselves. This development can be observed for the automobile industry but it is still not clear if it also applies for other industries.

Malone and his collaborators present a theory to explain this tendency of reduction of hierarchy. They argue that it requires an increased co-ordination effort to buy goods or services on the market because it is necessary to undertake certain time-consuming activities such as contacting suppliers or comparing prices. Co-ordination costs for hierarchical organisation are fairly low (a telephone call sometimes may be enough) but their production costs are much higher because of their monopolistic structure. Information technology increases efficiency in both kinds of organisations but this increase is larger for market oriented companies. Information technology reduces co-ordination costs considerably and by that stimulates market orientation. A typical example for this tendency is the just-in-time production in the automobile industry.

Malone et al (1987) describe some effects of information technology which are important in the context of their theory:

Electronic Communication Effect

The cost of information transfer decreases whereas the amount of information transfer increases.

Electronic Brokerage Effect

Basically, a broker is a person who provides contact between buyers and sellers. In this way he reduces co-ordination costs for both. Electronic brokerage transfers this activity to a computer systems. Electronic brokerage can reduce the cost of this mediation process considerably. This can result in an increase of the number of alternatives considered and therefore in an increase of the quality of the alternative that is eventually selected.

Electronic Integration Effect

This effect occurs when information transfer between two companies on adjacent steps in the value-added chain is not only accelerated but the information systems of both firms are integrated.

The main effect of the introduction of information and communication technology is an increase of the importance of the markets. But there are still areas where information technology supports hierarchical structures. CAD/CAM, electronic mail and other means can be used to intensify the co-ordination between the research and development department and the shop floor. These integration processes can include more than one company. On the other hand, information technology can stimulate new dependencies, especially in the context of the just-in-time strategy. There is a tendency to reduce the number of suppliers for parts (single sourcing) which increases the dependence between buyer and supplier as products get more and more specialised. Information systems are usually designed to support this process (Ofner & Pohl 1991).

Apparently, this development is contradictory. Malone & Rockart (1991) describe another hypothesis on trends in organisational structures. They argue that there is a tendency in companies to replace hierarchical structures by project groups which are supposed to carry out co-ordination tasks. They call this kind of organisation an "Adhocracy".

Adhocracy is a kind of activity which reflects the immediate necessities of a certain moment in time. This kind of activity is especially important in a very changeable environment where planning is almost impossible. Adhocracy also implies that such spontaneous decision processes are co-ordinated between several persons. Enterprises usually install project groups for a certain period of time. The groups consist only of those employees who are needed for the purpose at hand. Because of the high flexibility of the groups there is no tendency for stable hierarchies to develop (Malone & Rockart 1991, p.128).

Network systems are an important requirement for this kind of communication that is getting more and more important in modern business.

Conclusion

Recent research on the profitability assessment of computer technology shows that traditional methods of accounting cannot be applied exclusively. The utility of this technology cannot always be expressed in terms of money. The relationship between the source of a utility effect and its occurrence is very often not clear. Therefore, it is necessary to include qualitative aspects in the process of evaluation of information systems. One of these variables is the competitive position of a company. When deciding on strategic information systems managers usually also consider future trends in the market. Another important aspect is the organisation of the company that might be changed fundamentally in the course of the introduction of a network system. The example of the value-adding partnership shows that information technology, development of the competitive position and organisation are variables that cannot be investigated separately.

References

Biethahn, J., Muksch, H., Ruf, W.: *Ganzheitliches Informationsmanagement*. Band I: Grundlagen. München, Wien: R. Oldenburg Verlag 1990.

Clemons, E.C.: Evaluation of strategic investments in information technology. In: *Communications of the ACM*, January 1991, Vol.34, No.1, pp.22-36.

Dell'Mour, R., Fleissner, P., Müller, F., Ofner, F., Pohl, M., Polt, W.: Perspektiven von Technologie und Arbeitswelt II (Endbericht). *Projekt des Instituts für Sozioökonomische Entwicklungsforschung der Akademie der Wissenschaften.* Wien 1991.

Horvath, P.: CIM-Wirtschaftlichkeit aus "Controllersicht". In: H.H. Warnecke, H.-J.Bullinger (eds.): *Nutzen, Wirkungen, Kosten von CIM-Realisierung - 21.IPA-Arbeitstagung*. Berlin, Heidelberg, New York: Springer 1989, pp.131-148.

Johnston, R., Lawrence, P.: Beyond Vertical Integration - the Rise of the Value-Adding Partnership. In: *Harvard Business Review*, July-August 1988, pp.94-101.

Krommenacker, R.J.: The Impact of Information Technology on Trade Interdependence. In: M. Jussuwalla, T. Okuma, T.Araki: *Information Technology and Global Interdependence*. New York: Greenwood Press 1989, pp.124-136.

Malone, T.W., Yates, J., Benjamin, R.: Electronic Markets and Electronic Hierarchies. In: *Communications of the ACM*, June 1987, Vol. 30, No.6, pp.484-497.

Malone, T.W., Rockart, J.F.: Vernetzung und Management. In: *Spektrum der Wissenschaft*, November 11/1991, pp.122-130.

Moor, M.: Kostenstrukturen des CIM-Konzptes und Erwartungen (Nutzen) in CIM. In: H.J. Warnecke, H.-J. Bullinger (eds.): *Nutzen, Wirkungen, Kosten von CIM-Realisierung - 21.IPA-Arbeitstagung*. Berlin, Heidelberg, New York: Springer 1989, pp.291-308.

Ofner, F., Pohl, M.: Fallstudien. In: R.Dell'Mour, P.Fleissner, F.Müller, F.Ofner, M.Pohl, W.Polt: Perspektiven von Technologie und Arbeitswelt II (Endbericht). *Projekt des Instituts für Sozioökonomische Entwicklungsforschung der Akademie der Wissenschaften.* Wien 1991, pp.100-159.

Rada, J.: *Information Technology and Services*. Mimeo 1986.

Zangemeister, Ch.: *Nutzwertanalyse in der Systemtechnik*. 4.Auflage München 1976.

Zolles, K.: Theorie der Nutzwertmodelle und ihre Anwendung bei der Auswahl von elektronischen Datenverarbeitungsanlagen. (Diplomarbeit zur Erlangung des Grades Magister der Sozial-und Wirtschaftswissenschaften), Wien 1979.

7 Geographic Information Systems: Do They Support Business Objectives

GRAHAM ORANGE, LEEDS METROPOLITAN UNIVERSITY
PETER DIXON, TAYLOR WOODROW CONSTUCTION

Abstract

The researchers have been involved in developing a Mobile GIS for their own business use in field data capture. In liasing with clients they have also had the opportunity to explore the level of development, the different approaches, the differing perceptions and the general attitudes towards GIS in general. From these experiences many problems in the successful implementation of GIS into business particularly in the area of asset management have been identified. Drawing on their own experience and those of their clients the authors set out to identify the problems facing GIS in business and also to suggest that the development of GIS should be considered within the context of an information strategy plan.

Introduction

This paper has been produced as a result of our experiences in introducing Mobile Geographic Information Systems (GIS) to many varied organisations. The GIS has been designed primarily for use in the field for data capture and asset record maintenance arena (Dixon, Smallwood, Dixon, 1993) where data ownership is a critical aspect. Consequently our discussions with organisations have mainly encompassed those with distributed assets which have often needed a field survey to reliably determine their position (geocode) and attributes. Hence the need for mobile GIS.

Many situations have arisen where clients have identified, as the result of an information strategy planning exercise, the benefits they could achieve by having their information stored in a GIS. There are, however, many more situations where this enlightened view has not been apparent.

For Mobile GIS to be successful within an organisation it is necessary for us to identify clearly how a system is required to integrate within the business, the data and the functionality required. Experience has shown that value could be achieved by making use of facilities such as the graphical access and display options along with spatial querying facilities, which are commonly available in the more powerful GIS.

In this paper we have tried to identify the inhibitors which exist in the implementation of GIS, why these must be overcome and the ways in which this can be achieved.

Fundamentally we need to be clear what is meant by a GIS. Many definitions have been offered in diverse publications, the one, which we prefer, is proffered by Grimshaw (1994) who defines a GIS as:

> ...a group of procedures that provide data input, storage and retrieval, mapping and spatial analysis for both spatial and attribute data to support the decision making activities of the organisation.

The paper presented here refers to data as meaning asset data and not map data, in many instances in the data capture and maintenance arena, the map data is used for asset location and as an aid in the conceptualisation process. Frequently existing as little more than "wallpaper" to assist the human decision processes.

Why information systems need to satisfy business needs

We are in the information systems (IS) era. Information is being recognised as a major corporate resource and IT as the enabling mechanism (Earl, 1989). The IS era is replacing the data processing (DP) era though this migration to the IS era is not at a constant rate throughout all business sectors. Many organisations are still operating in the DP era and are technology driven. Information technology (IT) and its implementation needs to be planned and managed and, should be used to support business objectives (Earl 1989, Martin 1991, Orange and Orange 1994).

Systems should be developed to meet business requirements, preferably within the framework of a formal information systems strategy. All too often systems are developed to solve individual problem situations with little, other than coincidental, relevance to the objectives of the business. Application development is usually ad hoc and supports decision making only for a particular functional area. Such systems are often developed with a technical and operational bias. This leads to islands of automation where there is little integration (unless complex interfaces are built). Flexibility is compromised such that the organisation cannot respond quickly to the ever changing business environment. Organisations are faced with the problems of data redundancy and, more importantly, maintenance of data integrity.

Organisations which fall into this category are usually technology led. Systems are developed to fit the technical architecture rather than the other way around. Consequently the technology is assimilated into the organisation's existing procedures (Salzman and Rosenthal, 1993). The main objective of systems development carried out in this manner has been to reduce the clerical burden whilst benefiting from some of the advantages computerisation has to offer (e.g. speed and reliability). It's not the technology at fault but the approaches used to develop the systems.

> The problem lies not in the machine itself, but in the methods we use for creating applications - (Martin 1984).

Martin has since popularised user oriented strategic approaches which it is claimed produce systems that meet business objectives and satisfy user requirements. This is supported by Earl (1989);

> Firms need information (management) strategies which link the exploitation of information technology to business needs and also identify new business opportunities.

GIS within the information system context

To date, much emphasis in the design of GIS has been concerned with the geographic aspect of the data. This drive has meant that a great deal of importance has been placed on the G with little importance assigned to the IS. This is perhaps symptomatic of the fact that hitherto GIS has been the domain of Geographers. To support a business fully, we believe that the geographers need to be supported by Information Systems specialists. We maintain that it is in fact the information which satisfies business objectives with the geographical element providing additional power to the decision making processes.

This aspect is particularly relevant to the development of GIS as we consider GIS to be just another classification of information systems. There are a number of different classifications of information systems, for example management information systems (MIS), Financial Information Systems (FIS), Executive Information Systems (EIS). The discussion as to whether GIS should be treated as special applications is outside of the scope of this paper and is well argued by other authors e.g. Grimshaw (1994). However, the treatment of GIS in this way has worked particularly well within the Mobile GIS development strategy. The geographical aspect of the assets has been modelled as additional attribute data and its inclusion in the data and process models treated in this manner.

Instead of looking for new opportunities afforded by GIS there is the danger that GIS are used in the main as electronic mapping systems and not systems to provide management information. To do so is to not recognise the potential of GIS. It is symptomatic of the operational, technology driven approach. This is suggested by David J Grimshaw (1994) who states that there is little evidence to suggest that companies have moved into the business driven era of GIS development. However there is evidence to suggest that these perceptions are changing. Grimshaw (1994) makes the point that GIS whilst having been seen as a technology previously 'will in future be seen as a Geographical information system'. Advances in the enabling technology and the reduction in the costs of those technologies are facilitating the move from technical to business driven applications. I/S ANALYZER (1994) predicts that GIS are on the verge of becoming the "hottest business information tools." Similarly, ESRI in their White Paper Series (1992) concur:

> Geographic information systems (GIS) are becoming a mainstream data processing function in the 1990s. No longer solely relegated to solving engineering or scientific problems, GIS technology is being used in a wide variety of commercial applications.

However to realise the full potential of GIS, an earlier theme needs to be addressed. That is the requirement to consider GIS within the mainstream of information systems development. GIS should be developed to support the organisation's business objectives and satisfy its information requirements. To be successful, consideration must be given to how the GIS will integrate with other systems and to what extent it supports the organisational objectives and priorities.

The problem of the DP approach to systems development is that it creates islands of automation. When an organisation consequently looks at introducing GIS as a corporate tool then these islands of automation act as an inhibitor. The building of bridges (interfaces) becomes extremely complex and unmanageable. Typically this results in:

- Organisations going for a cheap solution to solve their immediate needs rather than the most cost beneficial solution for the organisation in general.

- Internal politics, rather than business requirements, dictating which systems are developed.
- Systems being funded from short term operational budgets rather than as part of a long term corporate plan.

Many management information systems contain geographic information.

It is said that about 80% of business data has some type of geographic component. - (I/S Analyzer 1994).

Moloney et al. (1993) suggest that this figure may be as high as 90%! When we consider that location information such as addresses and postcodes are contained in much of the data of many information systems then we can conclude then there is much commonality of data across an organisation's information systems.

The way that locational data is often defined using an unstructured DP approach is likely to be inconsistent. This will result in:

- Data redundancy and integrity problems.
- Systems designed without any analysis of data structure.
- Attributes specified without any thought about whether the data was required or used by the business, thoughts such as "we have always stored that" become common.
- Specifications based on procedural approach, typical of first generation systems, rather than using techniques that examine the underlying processes and data needs.
- Inflexible systems that cannot respond to changes in business objectives and priorities.

For an organisation's data to be viewed from a strategic level then commonality of data definition needs to be in place. Consequently, the introduction of GIS into organisations will highlight and exacerbate the problems associated with organisations grappling with the issues of strategic information system planning and the building of integrated systems.

If organisations simply take existing operational systems and convert them into GIS then it is not allowing the organisation to have the corporate view that GIS can provide. It does not allow for access to related data associated with particular assets held in other functional areas. For example, the cost of purchasing an asset may reside in accountancy and/or purchasing systems that are not integrated with the asset management system. Queries such as "the spend in a particular geographic area" are impossible without building complex interfaces between the isolated systems. Conversely, it does not allow for data held in different systems to be included in an overall geographic query. Queries about different assets within a given locational area are impossible without building further complex interfaces.

Grimshaw (1994) supports this by making the point that GIS are currently developed "to enhance the productivity or effectiveness of the functional area rather than to contribute to the business as a whole." It has been our experience that many GIS suffer from this and other classic symptoms of this operational approach resulting in a loss of credibility for the GIS and for the developers.

GIS are being caught up in the general information system development problem. The integration of GIS into an organisation with its Information Systems designed and implemented following a strategic planning exercise should find no difficulty in harnessing quickly and easily the full benefits of GIS. The overall success of this will of course be enhanced by updating the Information Strategy Planning exercise to include aspects of GIS in all the corporate models.

Perhaps some of the biggest obstacles to overcome are cultural. The take up of strategy planning approaches for MIS development has been slow. Is there any reason to suspect that acceptance of GIS within an information systems strategy will be any faster? To do so, GIS must be accepted by users and developers alike, as systems that support the business. For GIS to be acknowledged as more than a digital mapping tool it must be capable of being integrated into the existing information cultures. New procedures and technologies often involve change. Where change is implemented there may be resistance, a reluctance to accept new concepts.

> ...the introduction of GIS technology involves the complex process of managing change within environments which are typified by uncertainty, entrenched institutional procedures and individual staff members with conflicting personal motivations -. (Campbell 1992).

This is supported by our experiences. One of the biggest obstacles that has had to be overcome, is that of the users being dependant upon paper based systems for so long, along with their inbred familiarity with manual methods of operation. Likewise, users that do not appreciate the power that GIS have to offer perceive them as merely being digital mapping systems limited to reproducing the position of assets on a map.

We recognise the potential of GIS. The business driven approaches are offered to clients. When developing GIS for a client the emphasis is placed on the information requirements. Data and activity models of business information requirements are built using leading CASE (Computer Aided Systems Engineering) technologies. This facilitates compatibility and integration of GIS with the client organisation's existing information systems. Only when the client is satisfied that the information requirements will be met will the development of the technical interfaces be developed.

An organisation with no Information Strategy Plan is almost certain to struggle with business level GIS implementation. The unstructured "islands of automation" approach will not be successful. We believe that GIS is such a powerful tool at all levels of the organisation that its full value will only be achieved by a strategic implementation based on careful and wide reaching analysis.

In addition to the implementation of GIS the issue of data integrity is a key item which needs to be examined prior to the commencement of implementation. Our experience has shown that this is an absolute necessity. In many instances where data is remote from the location where the data is stored (as asset data is), the timescale for it becoming out of date is short it can be both difficult and expensive to maintain currency. Given the power of GIS to support strategic as well as operational decisions it would not be long before important decisions were being made based on out of date information. In some instances this might not be significant but in others it may be catastrophic. These risks must be examined and identified at a very early stage prior to implementation and suitable data maintenance facilities established.

As proffered above the method with most certainty of success in implementing a GIS into an organisation and achieving optimum value is to carry it out as part of a Strategic Information planning exercise. This may be either an update of an existing plan or the completion of a plan from scratch if necessary. We believe that this approach will ultimately prove to be inevitable. Those who carry out this exercise prior to commencing implementation will gain high rewards. Those who ignore it will do so at their peril.

Conclusions

The ongoing technical advances which are being achieved will open up a whole new arena for GIS. Hitherto GIS has manipulated historical data, soon it will be available to support dynamic business needs. The first beginnings of some of these systems are currently being seen in restricted functional areas. For this functionality to be harnessed to its full and accepted widely by business we believe that it is a fundamental requirement that GIS is implemented within business in a controlled and structured manner. In this way it will support the overall business objectives and not just the needs of an isolated department. If a single department can achieve good results with GIS, then business wide the impact can be substantial.

Many practitioners in the business worlds can visualise the powers of GIS and how it can support and enhance their organisations. We believe that it will not achieve these high expectations unless it is implemented in business in positive, structured and carefully planned ways. In some instances it will be necessary for organisations to "stand back" and examine their overall information strategy prior to attempting to implement and integrate a GIS into their main stream information systems. For those who do achieve this the rewards will be high, a case of the competitive use of IT. GIS technology is overtaking the perception of the average users and becoming the realm only of experts, business cannot be run solely by these experts. Geographical information must be available to all parts of the organisation for it to have a useful future.

References

Campbell H. Organisational and Managerial Issues in Using GIS. *Yearbook of the Association for Geographic Information 1992/3*.

Dixon P., Smallwood J. & Dixon M. Development of a Mobile GIS: Field Data Capture Using a Pen-based Notepad Computer System. *Association of Geographical Information Annual Conference*, Birmingham, UK. 1993.

Earl M.J. *Management Strategies for Information Technology*. Prentice-Hall. 1989

ESRI White Paper Series Dec 1992.

Grimshaw D.J., *Bringing Geographical Information Systems into Business*. Longmans Scientific and Technical. 1994.

I/S Analyzer, United Communications Group 1994.

Martin J., *Strategic Data Planning Methodologies*, Prentice Hall 1984.

Martin J., *Information Engineering Volume 1*, Prentice Hall 1991.

Moloney T., Lea A. C. & Kowalchuk C., *Manufacturing and Packaged Goods. In Profiting from a Geographical Information System*, GIS World Books Inc. 1993.

Orange G, Orange C, Strategic planning for Information Systems, *Informatik Forum*, December 1994, Band 8 Nr. 4.

Salzman H. and Rosenthal S. R., *Software by design: Shaping Technology and the Workplace*. Oxford University Press. 1993.

8 Strategies for the Introduction of Multimedia Systems

D J MOORE
LEEDS METROPOLITAN UNIVERSITY, LEEDS, ENGLAND
D J HOBBS
LEEDS METROPOLITAN UNIVERSITY, LEEDS, ENGLAND

Introduction

This chapter argues for the importance of applying strategic considerations to the utilisation of multimedia technologies, with particular reference to use of the latter as a training aid. It thus provides a concrete illustration of the practical importance of strategy as argued in other chapters in this volume. The value of multimedia technologies is first presented, followed by the case for employing a strategic approach if the potential offered by multimedia is to be realised. Finally some suggestions are made as to what such a strategic approach should comprise.

Why multimedia?

Multimedia will here be taken to mean the use by computer of more than text-only applications, in particular the addition of sound, video, graphics and animation. The related term "hypermedia" suggests a "node and link world with user-controlled traversal" (Shneiderman, 1992a). Such technologies offer many apparent advantages, especially from the point of view of education and training, as follows.

They can be expected to inherit the advantages of conventional computer based learning (cf. Mushrush, 1990; Steadman, 1994), and offer open- and student-centred learning opportunities. Thus learners are able to progress at their own pace, independent of time, place and teacher, and free of concerns about peer group pressure. Testing and monitoring of learners can be incorporated into the system, and the computer can offer the learner direct and immediate feedback, as well as appropriately tailored remedial work. Further, since training may be effected at the place of work rather than at a distant study centre, there may be cost benefits.

By employing a range of different media within a computer learning environment, further advantages may accrue. As Blattner and Dannenberg (1992) point out, the senses that humans use to interact with the world "enhance each other in various ways, adding synergies or further informational dimensions". By exploiting the senses, multimedia applications can potentially play a large role in enhancing and enriching learning. Multimedia documents, for example, can give clearer demonstrations via video and/or animation than their text-only equivalents (Zellweger, 1992), and a range of business training CD-ROMS are already on the market. In combination with international computer networks such as the World Wide Web, such innovations can potentially "empower teachers and students in remarkable ways" (Shneiderman, 1992b).

Further, by combining such facilities with a hypermedia facility, the system can take advantage of the analogy between the free movement between nodes and the associative process of creative thinking (Beeman et al , 1987). Learners are able to generate their own information-seeking strategies (Marchionini and Shneiderman, 1987), linking large bodies of knowledge to create representations that suit their particular needs, and browsing through learning material following and linking particular idea routes (Steadman 1994).

Much potential seems therefore to exist for multimedia applications within training. Further, training is itself a crucially important facet of modern working life (cf. Cumming 1988, Shneiderman 1992b). Parker (1995), for example, argues that pace of change is both "a threat and an opportunity" and, in a case study of the information systems division of a major British manufacturing company, found that staff were "reskilled" every four years. Thus technology is both causing the need for training, and, via multimedia, helping to provide means of providing such training. It is therefore vital that modern organisations consider how multimedia can help in their specific training needs.

As a result, perhaps, of arguments such as this, there has been much recent interest in the use of multimedia in training and education. The Computers in Teaching initiative (CTI) was established in 1986 to encourage computer assisted learning and teaching in UK universities, the University Funding Council have invested in projects involving hypermedia (Darby 1993), and HMI have identified seven areas of application of computers in the school curriculum (DES 1992). The private British company CRT Group PLC is donating multimedia computers to schools, to aid the development of the so-called "learning superhighway", in return for use of the rooms out of school hours as commercial training centres (Klein 1995). Currently the UK Engineering and Physical Sciences Research Council is seeking to fund research projects in the area of multimedia and networking applications with the expectation that they will "benefit not only HE but the whole of education, as well as training and conferencing in many other areas, helping the UK to become and remain more competitive". Furthermore, multimedia is already used in commercial training, for example in banks and building societies (Parker 1995), and in multimedia databases (Shneiderman 1992a) and marketing (e.g. Machover 1993).

Why a strategy for multimedia?

The case for organisational interest in multimedia is therefore very strong. It is the contention of this chapter, however, that a strategic approach must be taken to the introduction and utilisation of multimedia technologies if the benefits are to be realised in practice. Multimedia is not a sinecure for all problems - there are obstacles on the yellow brick road (Dickens and Sherwood-Edwards, 1993), which must be overcome to maximise the potential. It will not happen by right.

One such obstacle is the technology itself - there are limits to the technology's capabilities in a training role, in particular in regard to individualised feedback and discussion. In a study of a hypermedia package within a university, for example, Steadman (1994) found that learners keenly felt a lack of feedback from the machine compared with normal classroom feedback; one learner, for example, suggested "teachers can explain things in a variety of ways to aid understanding; computers can't". Similarly, in studying training applications in a commercial setting, Parker (1995) found that many learners felt that there could be no substitute for human interaction. Current work is addressing the issue of incorporating discussion into multimedia tools (e.g. Moore and Hobbs 1994) but this is a non-trivial issue, and it is difficult to disagree with Steadman's (1994) conclusions that, currently at least, technology is not a complete substitute for teachers, and technology-based teaching has an effective role to play only in particular circumstances and with regard to particular aspects of teaching. It may be therefore that there is a need for group use of training media, with the technology dovetailing into and thus enhancing, rather than replacing, conventional training. Indeed, there is evidence of its commercial use in such modes (Parker 1995). In brief, a strategic decision is needed vis a vis the interaction between multimedia technologies and conventional training approaches.

It is, moreover, important to realise that, partly because the area is a new and emerging one, many technical difficulties may need to be overcome if multimedia based training is to be successful. As Blattner and Danenberg (1992) point out, the use of multimedia systems tends to be poorly understood: "multimedia extensions to current systems have grown like weeds without well defined or well understood principles...We know we can add sound to existing systems, but how should sound be used? How does sound integrate with graphics? Do we really need sound for the particular application, or is it just adding meaningless bells and whistles?". A strategic approach is therefore needed to avoid the adoption of multimedia for its own sake. This concern is the more pressing since evidence concerning the effectiveness of CBL in general, and multimedia in particular, is hard to come by (Steadman 1994).

The opposite danger to the inappropriate adoption of the technology for its own sake also exists, namely the danger of "technological oversights" (Long and Long 1993). The field is developing rapidly, for example Fox (1994) has forecast 800 TV channels, with libraries of films available down phone lines and, as Vince (1993) has pointed out, the human longing to invent more and more esoteric technologies shows no limits. Even today, processors are under development that render obsolete virtually everything we are currently using. Some organisational strategy is therefore needed to maintain awareness of, and to cater for and take appropriate advantage of, the speed of development.

Other technical concerns are the presence of a number of different methodologies for multimedia development (Mulrooney 1995), and the presence of many authoring tools (see, eg, Bunzel and Morris 1994), suggesting the need for a strategic choice of development environment.

Alongside these technical concerns, are what might be called "softer" or more socially oriented concerns in connection with the implementation and deployment of multimedia systems. At the level of the individual learner, it is important that would-be trainees are aware of the possibilities multimedia training has to offer them and are assured that there will be opportunities to use the skills in existing and future roles (Parker 1995). Peppard (1990) has argued that open learning is a new environment distinct from traditional "classroom" techniques, and so requires strategic communication of the reasons for its use and of its benefits. Further, personnel may be wary of training using a machine and may require motivation and advice on how to exploit the benefits. Strategic planning is also likely to be needed on the part of the trainers. In his study of hypermedia in a university, Steadman (1994) found that lecturers needed time to become familiar with the computer based materials and to think through the implications of using them, and that their use required as much preparation and thought as any other teaching technique. He also found that there were differences between student expectations of the system and what it in practice provided them with, and argues that designers of such systems should be aware of and cater for the concomitant danger of "cognitive dissonance".

At the level of the organisation strategic decisions may need to be taken concerning the cost of the necessary hardware and software for multimedia training, and how it is to fit in with extant configurations (for example extant hardware may be capable of being upgraded to become full-specification multimedia machines). For an organisation producing multimedia training materials in-house there may be copyright issues concerning the rights to existing materials to be used in the package (Dickens and Sherwood-Edwards, 1993); for an organisation wishing to enter the multimedia industry as such, Dickens and Sherwood-Edwards argue for a need to think strategically in order to deal with the differing "fears and aspirations" of the various "players in the multimedia industry", such as rights owners, hardware manufacturers, and producers of materials.

More generally, two further considerations need to be taken into account at the level of the organisation. First, it can be argued that there are three levels of organisational management - strategic planning, management control and tactical planning, and operational planning and control (Hawryszkiewycz 1994). As with any information system, a strategic approach to multimedia should be taken in order to seek to deploy its benefits at each of these levels of management. Second, again as with all information systems, there are likely to be "political" ramifications of system implementation (Finnegan et al 1988): the system does not enter a value-neutral void. A strategic approach is therefore needed to manage the complexities of the system's introduction.

Towards a Multimedia Strategy

The argument is, therefore, that multimedia is a resource or tool an organisation should consider, but that its successful adoption is not something that can be taken for granted and must instead be carefully planned for. The attempt will be made in this section to provide some suggestions as to the likely nature of such strategic planning. The suggestions are necessarily tentative, given the emerging nature of the technologies since hypertext, hypermedia, and multimedia are all in the "Model T stage of development" (Shneiderman 1992a).

In order firstly to examine the contention presented above that successful implementation of multimedia training requires careful planning within the context of an overall strategy, Parker (1995) undertook two evaluative studies of the implementation of computer-based training systems in two large UK financial institutions. One of these delivered multimedia training based on the Phillips CDI technology; the other was a more conventional computer-based training package involving text and graphics.

The evaluation took place through user questionnaire and interview and was analysed against a set of twelve 'good practice' principles for design. These principles were developed by Parker on the basis of established HCI research in the context of training software and are shown below in Table 8.1.

1.	*know the user population*
2.	*allow query in depth*
3.	*design for user growth*
4.	*adapt to different user styles*
5.	*reduce cognitive load*
6.	*maintain consistency and clarity*
7.	*offer informative feedback*
8.	*utilise simulation*
9.	*aid perception*
10.	*provide navigational guidance*
11.	*indicate status*
12.	*use an effective pointing device*

Table 8.1:   **"Good practice" principles for design**

In comparing the two training systems, Parker concluded that they each had their own particular strengths and weaknesses. The CBT package was modular with clear aims and objectives at the module level. It included a 'bookmark' facility which enabled users to return to previously-encountered modules. Feedback was available through the 'review and discovery' questions and the test facility. The main perceived weaknesses were unimaginative graphics and the absence of sound.

The particular strengths of the CDI system were the benefits of full-motion video, high resolution colour, and audio capabilities which aided a high level of simulation and role play. A useful introductory option and modular aims and objectives supported a serialist learning approach, whilst a holistic aspect was provided by overview and in-depth supplementary sections.

Stage	Stage Title	Actions
Stage 1	Preparation	Discuss with the organisation the study objectives.
Stage 2	Where are we now?	Complete a SWOT (Strengths, Weaknesses, Opportunities, Threats) analysis to determine the current business situation. Discuss with the organisation the structure of the management hierarchy, their needs and the levels that may be affected by the proposed system.
Stage 3	Where do we want to be?	Discuss with the organisation the mission and the goals that they wish to pursue.
Stage 4	How do we get there?	Define functional strategies that will contribute to the corporate level strategy (mission). Establish what choices the organisation has and devise a set of options. Use a strategic planning tool to aid the decision making process.
Stage 5	Define the actions.	Establish whether the organisation should proceed with options devised in Stage 4, and choose which option to pursue. This may need further discussion with the organisation.
Stage 6	Establish the strategy/implementation plan.	Evaluate any other factors that may influence the plan - these will come out from the discussions with the organisation. Plan the implementation taking account of time scales and available resources.

Table 8.2:   Strategic planning guidelines

In summary, Parker suggests that CBT might be better suited to lengthy factual topics capable of segmentation whereas CDI might be better for business skills training requiring a stronger motivational element.

Parker next turned her attention to formulating strategic guidelines for incorporating multimedia as a training tool. Strategy is defined by Porter (1980) as 'a broad-based formula for how business is going to compete, what its goals should be, and what policies will be needed to carry out these goals'. Strategic planning does not follow a fixed sequence and can be an iterative process. However, Robson (1994) advocates that it is valuable to have a model of the planning process as a framework. Parker constructed a hybrid set of strategic guidelines derived predominantly from the Central Communications and Telecommunications Agency suggested guidelines for strategic planning of information technology, but including others advocated by Robson (1994) and Silk (1991). These are shown in Table 8.2 above.

A favourable appraisal from an independent expert helped provide validation of this set of guidelines. Application of the guidelines to the two companies mentioned earlier indicated that multimedia training would be beneficial to the training needs of both. Since one company had already made the decision some time previously to invest in multimedia and had already shown some level of competitive advantage as a result (see below), further support for the validity of the guidelines may be inferred.

Part of the application of the guidelines involved an analysis which concentrated on the functional level strategy of training in relation to the overall mission, and which used Silk's (1991) Benefit Level Matrix. This yielded suggestions that it would be cost effective for the company to consider the adoption of multimedia training and that further benefits would accrue from the development of in-house CD-ROM and CDI training software. The accuracy of this analysis was borne out by the fact that the cost-efficiency of multimedia had actually already been realised for the established user since the 'Motivating the Team' CDI training package cost £1175 to buy and set up for an unlimited number of trainees compared to £1425 per group of six employees for an existing training course run by an external training company (Parker 1995).

Conclusions

Multimedia systems generally, and multimedia training systems in particular, are currently experiencing enormous growth of interest and uptake. This chapter has argued that, as with conventional computer-based training in the past, lack of successful delivery of multimedia training may well result if the issues surrounding their adoption do not integrate within an overall strategic framework. A set of strategic guidelines has therefore been proposed as a first step towards assessing the appropriateness of such training and providing a vehicle for considered implementation. Follow-on research studies are now planned which will evaluate, refine and exploit these guidelines further.

References

Beeman W O et al (1987) Hypermedia and Pluralism: from lineal to nonlineal thinking, in *Proceedings of the Hypertext '87 Conference*, University of North Carolina,pp.67-88.

Blattner M M, Dannenberg R B (1992), *Multimedia Interface Design*, ACM Press (Frontier Series).

Bunzel M J and Morris S K (1994) *Multimedia Applications Development*, McGraw-Hill.

Cumming G (ed) (1988) Artificial Intelligence Applications to Learning and Training, Occasional paper In *TER/2/88*, available from ESRC-InTER Programme, Department of Psychology, University of Lancaster, LA1 4YF

Darby J (ed) (1993) The CTISS File, vol 15.

DES (1992) *Aspects of the Work of the Microelectronic Education Programme*, Report by Her Majesty's Inspectorate of Education, London, Department of Education and Science.

Dickens J and Sherwood-Edwards M (1993) Multimedia and Copyright: Some Obstacles on the Yellow Brick Road, in *Proceedings of Conference on Multimedia Systems and Applications*, Leeds, December 1993.

Finnegan R (1988) *Course units for O.U. course DT200, An Introduction to Information Technology*, Open University Press.

Fox D (1994) Manic Compression, in *Personal Computer World*, vol 17, no 10, p. 342.

Hawryszkiewycz (1994) *Introduction to Systems Analysis and Design*, Prentice Hall.

Klein R (1995) Link Learning - is it the business?, *in Times Educational Supplement*, March 17.

Long L, Long N (1993) *Computers*; Prentice Hall.

Machover C (1993) Multimedia Futures, *in Proceedings of Conference on Multimedia Systems and Applications*, Leeds, December 1993.

Marchionini G and Shneiderman B (1987) Finding Facts versus Browsing Knowledge, in *IEEE Computer* vol 21, no 1, p. 70-80.

Moore D J, Hobbs D J (1994) Towards an ITS for Educational Debate, in *Proceedings of the Interdisciplinary Workshop on Complex Learning in Computer Environments, Technology in School*, University, Work and Life-Long Education, Joensuu, Finland, 1994.

Mulrooney I (1995) The Development of an Interactive Multimedia Prospectus for Leeds Metropolitan University, Unpublished undergraduate thesis, Leeds Metropolitan University.

Mushrush J L (1990), Options in learning: Instructor Led and CBT, in *Journal of Library and Administration*, vol 12, no 2 p 47-56.

Parker A (1995) The Multimedia as a Business Training Tool, Unpublished undergraduate thesis, Leeds Metropolitan University.

Peppard J (1990) *IT Strategy for Business*, Pitman.

Porter M E (1980) *Competitive Strategy*, Macmillan.

Robson W (1994) *Strategic Management and Information Systems: An Integrated Approach*, Pitman.

Shneiderman B (1992a) *Designing the User Interface: Strategies for Effective Human-Computer Interaction*, Addison-Wesley.

Shneiderman B (1992b), Education by engagement and construction: A Strategic Education Initiative for a Multimedia Renewal of American Education; in Barrett E (ed) *Sociomedia: Multimedia, Hypermedia, and the Social Construction of Knowledge*, MIT Press.

Silk D (1991) *Planning IT - Creating an Information Management Strategy*, Butterworth Heineman Ltd.

Steadman C (1994) Some Studies on the Application of Commercial Software to the Production of Computer Aided Learning Materials, unpublished MSc thesis, Leeds Metropolitan University.

Vince J (1993) Introduction, in *Proceedings of Conference on Multimedia Systems and Applications*, Leeds, December 1993.

Zellweger P T (1992) Toward a Model for Active Multimedia Documents, in Blattner M M, Dannenberg R B (1992), *Multimedia Interface Design*, ACM Press (Frontier Series).

9 Organisational Integration of Expert Systems: Case Studies on the Phases of Design, Transfer, and Use

WERNER BEUSCHEL,
UNIVERSITY OF DORTMUND, COMPUTER SCIENCE DEPARTMENT,
DORTMUND, GERMANY

Abstract

This chapter reports on empirical findings from case studies in two large companies on the in-house deployment of small expert systems. The analysis focuses on the rationale of organisational practices during the stages of system design, field transfer and use. The results show the importance of integrative approaches to technical and organisational aspects of expert system development projects and methods.

Introduction

Not only technology matters in information system design, it is also the context in which design and use take place which is decisive for successful developments. The system being designed and its organisational context must be seen as interacting with each other, not as separate entities. To understand the results of system developments we have to focus on the process, not just on the product (Floyd 87). Thus, companies taking on system development are forced to deal with the interaction of system design and context, product and process. They thereby create their own practices. The notion rationale is used here to denote the underlying "theory" of practices. It comprises assumptions and strategies for decisions made during the life cycle (cf. Carroll 91).

Through the analysis of practices we may gain insights into characteristics of the development process. This should contribute to our understanding of design procedures as well as to the identification of "good" practices. The interactive perspective appears all the more valid for expert systems, as the necessities of soliciting, structuring and representing heuristic knowledge in many cases involve persons from other than the traditional programming departments from early on (Mumford and McDonald 89).

It was the goal of the two case studies to investigate how companies actually approach expert systems development, not aimed at research or educational purposes, but built for regular use. Practices are understood here as the set of strategies, activities and explicit reasoning applied during systems development with regard to technical, organisational and human factors. The analysis of the cases uncovers the underlying rationale in the practices and shows advantages and obstacles encountered during the development processes.

The cases were chosen from areas supposed to be characteristic, namely applications in offices, production and maintenance areas. In-house developments were selected as the organisational context of design, transfer and use is within the same organisational boundaries.

Methodology

With the growing distribution of information technology beyond its former confinement in data processing centers the insight grew that social and organisational aspects of system development processes become more important. Decisions in the course of these processes are neither just a matter of strategic choices of the management, nor of the capabilities of the technology alone. Rather it is the 'web' of organisational setting, available technology and the use of resources by the involved actors, which underlies the development process of computerisation (Kling and Scacchi 82). Expert systems can be seen as a special kind of interactive software. But as all information systems expert systems are amenable to a variety of functional and organisational design solutions, varying with the social logics inherent to organisations (Beuschel and Kling 93). They can facilitate augmentation as well as automation approaches. Therefore, the focus of the study was not on a pre-post-comparison of changes, but on the interaction of organisational, technical and human aspects of the development process and its consequences.

Reviewing the expert systems literature reveals that aspects of organisational implementation are not as well elaborated as technical characteristics (cf. Bernold and Hillenkamp 89, Bullinger and Kornwachs 90, Laswell 89). It was noticed that traditional design approaches are most often not sufficient in designing expert systems (CSS 89: 9). Proposals for new design approaches on the other hand are often not linked to organisational dimensions.

For the purpose of an empirical investigation it seems appropriate to state the selection criteria for systems more precisely. Two required features for selection were defined, as expert systems are introduced into work procedures where cognitive processes are involved: a distinct knowledge acquisition process during the course of the development process and the existence of a separate knowledge or rule base as part of the system infrastructure. The studies followed the way how a development process emerged within a company. Interviews and observations were conducted between 1989 - 1991. Depending on the operational status of a system three main stages of the process were analytically distinguished, the phases of design, field transfer and operational use.

Practices

The cases show practices for in-house-developments of expert systems in two large companies in California. One case depicts procedures to initiate the development of systems in all kinds of administrative or technical offices of an established computer manufacturing company in Palo Alto. The second gives the example of a diagnostic system for maintenance support in the field production of a worldwide operating oil company, its R&D department based in Southern California (for the full report cf. Beuschel 91).

EXOFFICE: Practices of a computer manufacturing company / technical and clerical offices

For obvious reasons the computer industry belongs to the earliest and most extensive users of its own products, since people are constantly encouraged and have easy computer access. This holds true also for AI-activities in the computer systems manufacturing company. A large and complex expert system was in the prototype stage to be used in wafer production of computer chips. But, according to one of the developers, it seemed to fail to represent the engineers' knowledge to the maintenance personnel and therefore was about to be redefined as a 'training system'.

At the same time, as a rather broadly aimed activity, the company decided in 1988 to promote in-house applications of expert systems on a less demanding level. A group was installed within the information technology department, here referred to as Advanced Systems Group

(ASG), with the aim of providing advising capacity on the automation of decision-oriented problems in offices. The group head was hired from within the company, with 10 years of experience in the hardware field of information technology, acquiring AI-knowledge by self-education. Both of his current collaborators in the small group were educated in computer science and cognitive psychology.

The special development strategy pursued in the latter case was to initiate the idea of introducing expert systems into the decision making of different departments. After receiving a request, ASG then essentially gave interested groups help for starting their own project. This was done by searching for appropriate applications and by providing tools. In some cases they also developed a prototype. But basically the department professionals were supposed to do their own knowledge acquisition with the help of the tools after a few start-up sessions with ASG. The rationale behind the approach was to reach "more consistency in decision making", as the ASG-manager put it, viewed as an equivalent to the idea of quality control in manufacturing.

In this way several initial projects were carried out, all of them but one being tested and in use since approximately fall '89. The average time for completion took about four weeks. While in these cases ASG did all the knowledge engineering, the expertise was provided by an expert at each division of the company.

Several problems showed up during the different stages of development. The group manager reported that it was difficult to correct unrealistic expectations of the departments concerning a match between technical capabilities, project volumes considered 'do-able' and the available knowledge about systems. Transferring systems from the design stage to regular use was considered a critical phase. It often only then showed up whether user involvement was appropriate or not. According to some examples, user interfaces tended to be too complicated and had to be simplified with regard to their information representation and interactional procedures.

The main remaining question was how to maintain the topicality and conformity of rules in the knowledge base. This was apparently neglected in the beginning by developers due to easy prototyping facilities. For long-term use of expert systems it was considered essential that the knowledge base had to be adaptable to new rules without allowing several copies of a system to develop in different directions. Individual users could not be allowed to have access. Thus the issue of developing organisational infrastructure was still to be resolved in all cases.

EXMAINT: Practices of an oil company / field production maintenance

Modelling and survey requirements motivated the use of voluminous data processing equipment in the oil industry during the past years. The potential of expert systems technology was tested from very early on. Examples of complex systems are Dipmeter Advisor, Mudman or Prospector, but the problems tackled were so severe that their use was confined to casual cases. Like the computer manufacturer the oil company commands a large department for research and development with approximately 600 engineers and computer scientists, also supporting all needs for data processing in all branches of the company. According to the analyst, soft- and even hardware was developed among several rival groups in the research departments. But in contrast, the company did not support separate AI-groups. All of those activities were based on personal interests and efforts of the researchers, sometimes in co-operation with universities.

The investigated example reports on a system built for a much simpler and frequently noted task for expert systems -- diagnosis -- in this case of wells, a part of oil production equipment. The system is in operation since 1987 with installations at 44 sites mostly in the US, but also one installation in Canada and in Indonesia. A site can be understood as an oilfield, containing about one to two thousand wells. Work groups are made up of a foreman, responsible for half of each field, with 2 or 3 production specialists for 200-300 wells and several pumpers, each one responsible for 60-70 wells. The research engineer estimates about 100-200 production workers as the regular users of the expert system.

In the past ten years several dedicated microprocessor based systems were developed by the research engineers to be used in all areas of oil field production. The systems also made it possible to narrow the job description of production specialists, from executing very broad tasks and responsibilities in the oil field to rather small maintenance tasks on the wells. As a (desired) consequence, personnel with lower skills had been hired since then. But afterwards there were delays in detecting and reporting maintenance problems, thereby reducing production capacity. So the idea for using an expert system was brought up by some of the research engineers. The development strategy was to use the available knowledge about the wells to build a system as part of the previously installed microcomputer system.

The system analyses data readily available from another program. It produces a diagnostic chart containing conclusions the production people may follow or use in their own way in order to regain the full production capacity.

Training offered in a 2-day-workshop was planned and introduced from the very beginning. It was primarily technical information on using the different systems but the research engineer stressed the side-effect for class participants of better understanding the organisational context of the company. Knowledge of procedures was necessary to enable the production specialists to report on system errors. It became clear from the training courses that it seemed easier for the production workers to accept advice from the expert system rather than from another person. They wanted to get advice on remaining questions they did not discuss during class sessions, since they did not want to show their ignorance there. The research department supported this activity, knowing that this not only would lead to feedback on system errors, but also to enhanced acceptance.

For the production workers it was rewarding to communicate with the research department. By using the protocol of the system output they had a more tangible basis to take precautions for maintenance. However their group heads, the foremen, weren't always convinced of diagnostics prepared with the help of the system. This gave reason to incorporate the foremen into the training plan as well. So the training course acquired an important function as a 'missing link' in the communication process between different organisational units.

The rationale

Two salient strategies connect the reported cases. The first is the use of the expert system approach for small and surveyable tasks rather than for tackling complex and large problems. The second is the managerial emphasis put on organisational as well as on technical integration and not just the installation of a new computer system. Both strategies and their implications shall be discussed in some detail.

In both companies the characteristics of tasks viewed to be appropriate for an expert system approach were similar. The tasks in question were part of prestructured work procedures, involving heuristic knowledge about situational or temporal circumstances and making use of more or less intensive numerical calculations in various cases. Only part of the heuristic knowledge was tried to be incorporated into the system, which may be a factor for the relatively quick and successful implementations. This differs strongly from tackling ill-structured tasks in expert system developments, which was subject of research efforts in many experimental examples. It seems that the simplified approach is meanwhile more common among business applications.

While the strategy followed in EXMAINT keeps firmly an eye on compatibility and integration of the expert system with regard to the existing microprocessor systems, the characteristics of the intended user group are continually reflected from the very start, and so plans for organisational aspects are made at the same time the technical system is developed. In contrast, the emphasis in EXOFFICE is on organisational participation, in order to avoid too much friction between development group and intended user departments. Here, the choice of technical tools clearly follows this premise. Regardless of the different emphasis, both cases pursue an integrative approach of technological and organisational aspects.

The following collection denotes the rationale for development practices, on the basis of common characteristics of the two cases:

Design:

- Design with the available hard and software equipment in mind.
- Select simple tasks, requiring not too much personal experience, avoiding ill- structured tasks.
- Involve users as early as possible.
- Choose dedicated AI-tools only if no simpler tools will suffice.
- Keep development project, system and knowledge base as surveyable as possible.
- Design the user interface from the user's view, not the designer's.

Transfer:

- Define the final functionality of a system only after going back and forth between design and transfer phase several times (iterative design).
- Plan for field tests and system revisions, since introducing the system into the application field is not a simple 'installation'.
- Reduce the complexity of user interfaces to what seems appropriate for users of the application area.
- Functional solutions for a task can be provided to non-experts as well as for experts.

Use:

- Hands-on-experience for users is indisputable, may it be acquired by participating in the design process or by training.
- Maintenance and update tasks gain importance with the number of applications.
- Consistency of the knowledge base is crucial for multi-site applications, though no organisational procedures are readily available.
- Developer(s) and expert(s) must be in reach until a system has gained its 'stable state.'
- Non-experts may possibly increase their competency by using the new system.
- Normalisation of work procedures allows for more precise or narrowed job descriptions.
- Growing degree of formalisation points to possible future forms of organisational and technical integration.

Conclusions

Practices for design, transfer and use of small expert systems in different industrial areas were analysed. The examples showed the emphasis put on technical and organisational integration by the developers. But despite this approach unforeseen outcomes of the development processes still occurred, though on a small scale.

What becomes visible behind the rationale is that even small and unobtrusive expert systems require a constant focus of attention on organisational issues, especially during the early stages of the life cycle. This might be invoked by technical aspects (like new user interfaces), social aspects (like hierarchical frictions due to small changes in the availability of decision-making support), or infrastructural aspects (like available training facilities). So the question remains, to which degree this kind of integrative approach to expert systems development, as characterised above, could hope to avoid all organisational obstacles. Though integrative approaches like the ones described here can be improved gradually by learning from previous pitfalls, they are nevertheless likely to encounter non-planned events during future projects. Thus the most important aspect is to prepare for development as an ongoing, open process.

The important practical lesson to be learned from these and other examples is that the more human knowledge -- and not just number crunching -- is the subject of computerisation, the more organisational efforts will be necessary in order to reach the goal of a usable system. The cases point to the limits of development practices based on an idealising theory of human knowledge. Contradicting an often heard belief, the cases also speak for the continued necessity of training efforts in the context of expert system developments.

References

Bernold, T.; Hillenkamp, U. (eds.): Expert Systems in Production and Services II. From Assessment to Action? *Proceedings of the International Workshop on Expert Systems,* Chicago, USA, September 13-15,1988, Elsevier: Amsterdam 1989.

Beuschel, W.: The Rationale of Development Practices for Expert Systems - An Empirical Investigation. Technical Report No. 91-22, *Information and Computer Science,* University of California, Irvine.

Beuschel, W.; Kling, R.: Computerisation and Workplace Transformations. In Luczak, H., A. Cakir, G. Cakir (eds.): *Work with display units,* Springer: Berlin 1993: 395-399.

Bullinger, H.-J.; Kornwachs, K.:Expertensysteme *-Anwendungen und Auswirkungen im Produktionsbetrieb.* Beck: Muenchen 1990.

Carroll, J. (ed.): Special issue on design rationale. *Human Computer Interaction,* 6(1991): 3-4.

CSS (Council for Science and Society, ed*.): Benefits and Risks of Knowledge-Based Systems.* Oxford University Press: New York 1989.

Floyd, C.: Outline of a paradigm change in software engineering. In: Bjerknes, G.; P. Ehn and M. Kyng: *Computers and Democracy,* Avebury: Aldershot, pp. 191-209.

Kling, R.; Scacchi, W.: The Web of Computing: Computer Technology as Social Organisation. Advances in *Computers,* vol. 21(1982): 1-90.

Laswell, L. K.: Collision: Theory vs. Reality in *Expert Systems.* QED Information Sciences: Wellesley/MA (2nd ed.) 1989.

Mumford, E.; MacDonald, J.B.: *XSEL's progress: the continuing journey of an expert system,* Wiley: New York 1989.

PART III
USER PARTICIPATION

10 Participatory Standardisation of Information Infrastructure

OLE HANSETH,
NORWEGIAN COMPUTING CENTRE, BOX 114 BLINDERN, 0314 OSLO,
NORWAY
ERIC MONTEIRO,
DEPT. OF INFORMATICS, NORWEGIAN UNIV. OF SCIENCE AND
TECHNOLOGY, 7055 DRAGVOLL, NORWAY

Introduction

Development of and use of information infrastructure (II) is receiving much attention these days (Kahin and Abbate 1995). There are numerous issues at stake in establishing a working II including: the prospects of promoting more adaptive and flexible inter-organisational collaboration (Davidow 1992), redefining the role of governmental intervention (Kahin and Abbate 1995; Mulgan 1991), balancing the self-reinforcing effects of further diffusion of the II with the need for flexible changes (Antonelli 1992; Hanseth, Monteiro and Hatling 1996) and understanding the specific problems of adoption and use of IIs (Star and Ruhleder 1996). This paper focus on the issue of participation from relevant users in the development and appropriation of IIs. More specifically, we inquire into the underlying rationale for participatory design (PD) in general and discuss how to translate this to IIs. This involves identifying the socio-technical character of the obstacles of actually achieving user participation and influence in II development.

The remainder of the paper is organised as follows. In section 2 the notion of participatory design is sketched. The idea is to make explicit some of the underlying motivation for participation in the development of IIs2E Section 3 outlines relevant aspects of the process and organisation of IIs within health care. The illustrations from the case is contained in section 4 and concluding remarks are provided in section 5.

The paper is based on illustrations from a case of II within the Norwegian health care sector.

Methodologically, the case is an historical reconstruction from written records (minutes, reports and standards) together with a handful of unstructured interviews with involved actors. The case is described in much greater detail elsewhere (Hanseth and Monteiro forthcoming).

Participation in technological development

Politics or Pragmatism

The underlying rationale for participation in the development of technology vary. There is no single, overriding motivation. It is rather so that different kinds of arguments and motives are combined and superimposed. It is relevant to sketch, at least briefly, these motives. To simplify, the different motives and arguments for participation come in every shade from a political or ideological one to an utterly pragmatic one (Mumford 1984; Winner 1992; Miller 1992). The political or ideological argument for participation is linked to concerns for democratic control and quality of work life. The argument, in brief, is this. Democratic control is an universally recognised goal. As people spend a significant amount of their time at work, the principle of democratic control extends to the realm of working life also. When technological development of artefacts used at work influence the quality of work life, their development should likewise be subject to the principle of democratic control (Sclove 1992).

The pragmatic reasons for a concern for participation is linked to traditional concerns for cost-effectiveness (Schuler and Namioka 1992). If the participation of (various categories of) users is a cost-effective way of designing technology, why not use it? (Mumford 1984hvor). A number of empirical studies exist which attempt to document this positive effect of participation. But the empirical evidence is not conclusive (Baroudi, Olson and Ives 1986). Note that the identification of the different arguments for participation needs to be considered a purely analytical device. The continued interest in participatory design (or whatever one might like to call it) is sustained exactly because these different arguments in a given project exist in conjunction. There is accordingly a sliding difference between a focus on pragmatic issues (over cost-effectiveness) and viewing the pragmatic concerns as means to achieve more political or ideological goals.

Economic Theory of Technological Development

Within economic theory there is a growing appreciation for evolutionary development of technology with users playing an important role (Antonelli 1992). Technological development is here conceived of broadly, encompassing innovation, adoption and diffusion.

The basis of this thinking is the recognition that technology is not to be treated as a "black box". Ideally, the designers learn about the need of users while the users learn-by-doing, that is, extend their capabilities and understanding through practical use. Hence, technological development should be recast along the lines of enhancing an evolutionary process of learning at the consumption as well as the production side of technological development.

Large Technical Systems and the Need for Standards

The development of large, complex, geographically and organisationally dispersed technology like an II pose a number of challenges for the role and extent of participation. We identify a few of these.

The point made above about the importance of enabling a stepwise learning process alongside the technological development is more difficult for IIs. This is due to the fact that the way situated, contextual learning-by-doing is to scale and accumulate to the II as a whole is anything but clear. This presupposes a strategy for sustaining local flexibility over long periods (Hanseth, Monteiro and Hatling 1996). IIs are to function as a common communicative infrastructure among a diverse collection of user groups or organisations. This immediately creates difficulties in reaching agreement over the underlying standards. The different user groups may, and typically will, have different interests, conceptions and roles in the work processes associated with the use of the II. The multiplicity of interests is difficult to concert in the absence of a central authority (Neumann and Star 1996). If the development of the II is not totally dominated by a single actor (organisation or user group), reaching consensus among the actors need to be institutionalised in some form. Several possibilities exist include voluntary boards, consortia or formal standardisation bodies.

The Invisibility of II

There is a paradox related to the involvement of users in the development of II (Neumann and Star 1996). It is difficult to mobilise interest into the design because a working II is to a large extent transparent. It only becomes visible in break-downs.

The complexity - in terms of the number, type and size of the technical components together with the social and bureaucratic fabric of the standardisation process - make it practically impossible for any one of the actors involved to grasp the artefact as a whole (Ciborra 1996). As a result, it becomes quite oblique how the different elements of the II inscribe behaviour from the others (Hanseth and Monteiro forthcoming).

Experience with Developing IIs

Future IIs evolve by combining, extending and aligning existing IIs. Existing application level IIs may be grouped into three broad categories:

- tailor-made, designated IIs, for instance, flight reservation and ban
- king networks;
- exchange of form like messages and EDI, for instance, within car industry, finance and dealing with subcontractors;
- general purpose IIs with applications, notably Internet and OSI;

The various issues in relation to participation outlined earlier are handled in different ways in each of these cases. No systematic evaluation exist, to the best of our knowledge so our comments are collected from different sources.

The development of specialised IIs, for instance, flight reservation, has taken place in consortia with dominating key actors. The problem is reasonably well understood among the different actors. The use of EDI has, despite some exceptions, been disappointingly slow in diffusing. This is especially true for EDIFACT standards (Graham et al. 1995). As the development of EDI messages gets injected into the EDIFACT machinery, participation from users become practically impossible (Hanseth and Monteiro forthcoming). The only general purpose II in widespread use is Internet. Participation in the ongoing design of Internet is in principle open (Hanseth, Monteiro and Hatling 1996) - but not in practice. Traditionally only a fairly small number of (largely US based) designers are capable of knowing enough about the jargon, culture, technology and process to actually be able to influence the design of II. With the increasing importance of industrial consortia developing Internet technology (primarily web-related technology and electronic payment), this picture is changing.

Standardisation processes

One normally distinguishes between de facto, de jure and formal processes of standardisation. De facto standardisation is characterised by its reliance on market forces; there are no regulating, institutional arrangements influencing the process. De jure standardisation denotes the situation, typically within a hierarchical organisation, where standards are approved by one, central organisation, for instance, a suitable, national governmental institution. The third type of process, formal standardisation, is most relevant in the present context.

II, like many other kinds of large technical systems, are standardised by formal, quasi-democratic, international standardisation bodies (Lehr 1992). These standardisation bodies have to respect predefined procedures and rules regulating the status, organisation and process of developing standards. In recognition of the limits of both market forces and hierarchical control, formal standardisation is a key strategy for developing an II 2E. There are three important institutions responsible for formal standardisation of II:

- the International Standardisation Organisation, ISO (and its European branch, CEN);
- EDIFACT within the United Nations;
- Internet;

These three institutions organise the process of standardisation quite differently along several important dimensions including: the way participation in the process is regulated, how voting procedures are organised, the requirements proposed standards have to meet at different stages in the process, the manner information about ongoing standardisation is made public and the bureaucratic arrangements of how work on one, specific standard is aligned with other efforts. For a more detailed description of these differences, consult (Graham et al. 1995; Hanseth, Monteiro and Hatling 1996; Lehr 1992).

De facto standards are developed by industrial consortia or vendors. Examples of such standards are the W3 consortium currently developing a new version of the HTML format for World Wide Web, IBMB4s SNA protocol and the HL-7 standard for health care communication.

Health II is use of an II within the health care sector. It has evolved over a period of ten years and takes different shapes over time. Its two main types are transmitting of form-like information and (possibly real-time) multi-media information. Illustrations of the former include: lab orders and reports exchanged between general practitioners (GPs), hospitals or labs and (other) labs, admission and discharge letters between GPs, specialists, and hospitals, prescriptions from GPs to pharmacies, exchange of non-medical information like ordering of equipment and food and invoices from hospitals and GPs to health insurance offices for reimbursement. Illustrations of the latter type include: telemedicine services, that is, computer based services which usually include real time multi-media conferencing systems supporting a physician requesting advise from another physician at another institution, access to data bases and Web servers containing medical information and PACS (picture achieve systems for X-rays) systems.

Standardisation of health care information exchange has been going on for years within numerous bodies in conjunction with various collaborative and competitive relationships. An early standardisation effort started in the US in the mid-80s through the establishment of the HL-7 consortium. This consortium was established by small vendors in order to develop a shared standard as an important tool to strengthen their competitive position compared to the larger vendors. Accordingly, they did not allow the latter to participate. The resulting HL-7 standard has been actively promoted in Norway and other European countries by Andersen Consulting. At the MEDINFO conference in 1986, an initiative to develop open, international standards was taken. This resulted in the establishment of the IEEE 1157 committee, usually called Medix. The Medix work was considered the most important effort among those believing in open standards until 1990, when the Commission of the European Community (CEC) asked CEN, the European branch of ISO, to take responsibility for working out European standards for information exchange within the health care domain. For CEC, standards were considered important vehicles for establishing the European inner market. With the financial and political support of CEC, CEN has become the most powerful standardisation body in the health care domain (CEN1993c). The CEN work has been co-ordinated with health care related standardisation work in other parts of the world (US, Oceania and South-East Asia). When CEN in 1991 decided to base the development of lab messages on EDIFACT, parts of their work was delegated to Western European EDIFACT Board (WE/EB), a subsidiary of the United Nations EDIFACT Board. CEN and WE/EB established a liaison agreement covering those health messages which were based on EDIFACT.

A case from health care

Background

Our focus in this paper is on the development of Norwegian standards for exchange of lab orders and reports together with drug prescriptions. This standardisation work is tightly intertwined with international standardisation efforts on various levels and within various application domains. Standardisation of health care information exchange has been going on for years within various bodies in conjunction with various collaborative and competitive relations. In 1990, the Commission of the European Community (CEC) asked CEN, the European branch of the International Standardisation Organisation (ISO), to take on the responsibility for working out European standards for information exchange within the health care domain. For CEC, standards are important tools for establishing the European inner market. With the financial and political support of CEC, CEN has become the most powerful standardisation body in the health care domain. Exchange of lab information is a rather complex issue. First, there are different kinds of labs including clinical-chemical, micro-biology and pathology. Moreover, the information is exchanged between a wide range of units: between the various instruments and a central database within the lab, between the lab and test ordering units such as hospital departments, other labs, and General Practitioners (GPs). A project team within CEN is working on a coherent model for the whole lab field and taking care of the integration with the standardisation of information exchange in other areas within health care. Lab orders and results will be exchange between GPs and labs as EDIFACT messages. This is also true for the transmission of drug prescription from GPs to pharmacies. CEN has delegated the standardisation of these and other EDIFACT messages to Message Development group 9 (MD9) within the Western European EDIFACT Board (WE/ EB). WE/ EB is a part of the world-wide EDIFACT organisation under United Nations. The Norwegian Centre for Medical Informatics (KITH) co-ordinates national standardisation activities in Norway and their relations to international ones. In Norway, the standardisation work includes tasks like specifying Norwegian requirements to a European standard, validation of proposed European standard messages and to specialise European standardised messages according to Norwegian needs.

The standardisation of lab and drug prescription information is organised as projects. We will here give a more detailed description of the drug prescription project only. It is responsible for the development of a functional- and implementation guide for drug prescription as well as to co-ordinate a concrete implementation for pharmacies and GP applications.

In drug prescription there are several interested parties. The key participants are: the pharmacy professionals, the GPs, and the public health insurance (paying for a large part of the drugs). Moving from paper based to electronic drug prescriptions requires the Cupertino of all the involved parties as the GP cannot transfer any electronic drug prescriptions unless the pharmacies can receive them. And the health insurance depend on electronic transmission from the pharmacies. But all of them do not have equally obvious benefits from co-operating. Both pharmacies and the health insurance are very motivated and interested in the project as it makes their work significantly easier. The pharmacies find that it saves work because they do not have to retype the drug prescription into their own application any longer. The health insurance see it as a means towards more effective cost containment. The health insurance pay for approximately 50% of the total use of drugs in Norway, the patients pay for 32% and the hospitals 18%. The GPs, however, do not have the same kind of direct advantages in the project. To make them participate in the project, they must be offered other benefits. The solution is to attempt to support the work of GPs by providing them with better information about drugs and improving the quality of drug prescription by eliminating errors.

The drug prescription project started as a pre-project. The participants in the pre-project were representatives from:

- the association of GPs;
- he association of pharmacies;
- KITH;

The associations of the different professional groups at the pharmacies were not involved in the pre-project. New working routines at both the GP's office and the pharmacy were identified, but no solutions were discussed or prepared (KITH 92). This was delegated to later phases of the project. The main project is organised on two levels, a co-ordination and a project group. In the co-ordination group there are representatives from the same organisations as in the pre-project together with:

- the professionals in the pharmacy;
- the health authorities: the payers and those who develop legislation/instructions;
- national EDIFACT-organisation (Norwegian TEDIS);

The co-ordination group keeps an eye on the project without making any decisions. The purpose of the group is to function as an arena for discussion. The co-ordination group has met every fourth month or whenever necessary.

In the project group we find representatives from:

- the suppliers of applications for GPs;
- the (only) supplier of applications for pharmacies in Norway;
- the supplier of an EDIFACT converter;
- the national association of GPs;
- KITH;

The project team has the overall responsibility for the project. They are responsible for the development of the implementations guide, specification of pharmacy and GP systems, and the final evaluation of the project. The project team has met approximately once a month. The project team has mainly been concerned with requirements specification and security aspects.

At the present, the suppliers are working with the implementation in internal groups. Such internal groups are found at every supplier. The co-ordinator has little influence over these internal groups.

In the main project, the focus has been on the technical implementation of the drug prescription message in the GPs' and pharmacies' computer systems. The requirements specifications for the GPs' application (Carlsen 93) and the pharmacies' application (NAF-Data 94) deal with new working routines. These include issues which have been discussed but not yet settled 2E The final decisions are delayed until some experience is gained. Examples of such issues are:

- What will the patient do if she goes to a different pharmacy than the one specified by the GP?
- Does the patient has to go to the same pharmacy to get drugs from a reiterated prescription?

The health insurance plan a future computer system which will be affected by the drug prescription project. When their future computer system will be implemented, the pharmacies and the GPs have to change and co-ordinate their routines with the health insurance's new routines.

The drug prescription project has chosen an object oriented information model which is subsequently mapped to an EDIFACT message (KITH 93a). This approach is also chosen in other health care standardisation activities, nationally and internationally. Both areas (the model and the mappings) require a reader with some technical skills. The health care users have clinical backgrounds, and are in very few cases technically skilled. This becomes evident from the comments from the hearing after publishing the previous implementation guide for laboratory reports in Norway in 1991 (KITH 91). Most of the comments from the clinical personnel expressed the view that the document was too technical. Few organisations were able to comment both on the information model and the EDIFACT-syntax of the message.

Identification of Prescribed Drug Item

The issues we hope to illustrate by this example are basically: how an II for electronic exchange of drug prescriptions affects existing work practices and applications, how a seemingly technical design issue is translated into non-technical issues, and how one tries to develop the system so that everyone benefits from it. In addition, it also shows how the actors make plans for external and non-participating parties. Exactly who the involved parties were was not clear from the outset, it became evident only later.

Early in the project the pharmacies expressed their wish to use the prescribed drug identification number in the electronic drug prescription message. This piece of information was not available in the paper version of the prescription; the electronic version created a new opportunity. This number uniquely identifies the drug prescribed by the GP. The pharmacies wanted these identification numbers because they could subsequently be transmitted to their existing application for electronic ordering from The Norwegian Drug Medication Depot (NMD). In this way they would achieve an explicitly stated goal: improved logistics (Statskonsult 92).

None objected to the suggestion to include this number in the message. All agreed because this information did not seem to conflict with anyone's interest. The problem was only that the GPs made no use of it and hence had no access to it in their work. Neither the GPs nor their applications were aware of which identifiers corresponded to which drugs. The initially straightforward and technical issue of how to code and represent a prescribed drug identification number in the message had thus been translated into another issue: How should one find a way for the GPs to provide this number as part of the prescription without creating additional work for the GPs? The GPs do not need to see this item number, and the application could deal with it non-transparent to the GP. Two sources for obtaining the information were considered: the drug item list and the Common Catalogue (Felleskatalogen).

The pharmacies' drug item list is provided by the NMD, and they make it available to the pharmacies via their application supplier. The list contains information useful also for the GPs, e.g. about price and synonymous drugs. The GPs' representative argued that this list had to be made available for the GPs' applications. The drug item list is updated every month, and would definitely improve the prescription work of the GPs. Today the GPs use their own drug lists which are either typed in by themselves or installed with the application (as a supplier's service). In either case, the GPs themselves have to update their drug list.

As everyone agreed that the GPs need the drug item list, the focus has been on whether or not the GPs should pay for access to the drug item list, and if so at which price. The Pharmacies' Association has recognised the GPs' need for the list. They have no principal objections against offering relevant parts of it to the GPs. The NMD, however, who is responsible for the drug item list, has not yet decided what to do. NMD is not represented in the project and will probably not offer the list free of charge. The GPs cannot get access to the drug item list as it is at the present, because the list also contains information which the pharmacies want to keep for themselves (e.g. about profit margins on pharmaceutical products). The list thus has to be tailored to the needs of the GPs. This could either be integrated with the drug list in the application or a complete new list could be composed. This raises the question about who should do this job, and who should distribute the list? At the present, it seems like the NMD and the supplier of the pharmacies application are trying to find a solution to this problem. The most likely one seems to be that the NMD will offer the list to the supplier of the pharmacies application. It has not yet been decided whether the NMD or the Pharmacies' Association should adjust the drug item list. The supplier of the pharmacies application will subsequently distribute the adjusted drug item list to the different suppliers of GPs medical record applications. And these suppliers will pass on the list to their customers. The drug item list will probably also have to be known by the health insurance, to enable them to check the prices used by the pharmacies.

The GPs also have available a paper based catalogue, called the Common Catalogue, containing information about all registered drugs in Norway. It also contains other important and practical information about treatment of acute poisoning, drug interactions, a register of drug producers and a register of all pharmacies in Norway. But this catalogue is provided by yet another organisation, an organisation not part of the drug prescription project. The catalogue is printed once a year, but additions are printed during the year as new drugs are introduced or disappear. The Common Catalogue now exists electronically as well. Work is being done to integrate the catalogue with the GPs' drug list to ease the work for the GPs. This is seen as a possible solution to the problem above, i.e. how to offer something to the GPs which makes it acceptable for them to register the desired prescribed drug identification number.

This integration work requires the Cupertino of GPs, suppliers of GP systems, the NMD and the organisation responsible for the Common Catalogue. All these parties have commercial interests in this area, and their motivation must be combined with GPs' actual need for integration of drug information. This work has just started, and in the mean time the drug prescription exchange project has to solve the problem with how to make the item prescription number available for the GPs.

Necessary Identification of Parts of Laboratory Messages

This example is primarily intended to show the impossibility of developing an II without getting heavily involved in both existing applications and work practices.

The new European standard for laboratory messages is based on the following scenario (CEN 93a): Usually the GP sends an order (with a physical specimen) to the laboratory. The laboratory analyses the specimen and returns a laboratory report to the GP. The patient has to pay a part of the expenses, and the health insurance pay the rest. If the requested laboratory cannot perform the necessary analyses, it can transfer the order and the specimen to yet another laboratory. This second laboratory returns its report to the first laboratory, which merges the two partial reports and returns a complete one to the GP. To do this, orders and reports must be uniquely identified. And all involved parties need to know the use and meaning of identifiers. Electronic exchange of drug prescriptions and laboratory order and report messages requires unique identification of:

- the health care professionals (GPs, laboratories);
- patients;
- orders and reports;
- specimen;

We will here take a closer look only at the identification of laboratory orders by order numbers. Both the sending and the receiving applications need order identifications. Today this identification is not generated in a standardised way. Every application has its own identifiers. This implies that the suppliers need detailed information about the use of order numbers both in the sending and receiving application. In the case of mistakes or errors, both clinical and technical personnel need to refer to the right order. The order number also appears on the bill the patients receive. If the order numbers are used for additional, local applications, say within economic modules or transferred to the payer, this number needs to be known to all the involved parties. One thus has to co-ordinate this unique identification with other, surrounding applications and work routines.

Most of the electronic medical record applications for GPs in Norway use their own internal record number to identify each patient's journal. They do not use the person's social security number. Currently they have no mechanism for unique identification of orders and specimen needed to link them together. They rely on the assumption that a combination of the patient and the specimen collection date is sufficient. This requires that a GP office never sends more than one order on the same specimen on the same day for any given patient. A change of this requirement would imply a restructuring of the application. We thus have a situation where a technical issue, representation of identifiers, has been translated into a non-technical issues, strong assumptions upon the work and behavioural patterns of the GPs and their patients.

The order numbers used at the laboratory is an unique identification of the order and patient. But few of the GPs' applications are capable of using or storing this information. Each laboratory has their own order number series, and even different laboratories within the same hospital use different order number series. GPs transferring electronic orders to different laboratories, each with their own system for unique representation, would need to keep track of all these odd systems. Also laboratory to laboratory communication would be complicated with so many different requisition number series.

Economic Information

This final example is intended to show how the actors protect their interests by including information they desire through active participation, i.e. translate abstract interests into specific design. It also illustrates the flexibility of standardisation as not all interests may be expressed within the message itself; they have to be taken care of in the system architecture. Finally, it shows the new possibilities for services which the electronic medium creates.

Both in the case of laboratory tests and a certain kind of prescriptions, the health insurance pay for a part of these services. As the costs of these services are significant, they are quite understandably concerned with effective cost containment strategies. In connection with the two standardisation processes we study, this surfaces as a concern for including appropriate information which they may use for these purposes. Here we only consider the case of drug prescriptions.

At the moment, the health insurance receive a huge paper list with documented expenses from the pharmacies. The health insurance are supposed to check these lists, but in practice they have no possibility of doing this in a cost-effective way. A new computer system will be installed within the health insurance during this and the next year. This system will make it easier to control what is prescribed to whom. The electronic exchange of prescriptions, each having a unique prescription number, will make it easier for the local health insurance to carry out this control. The health insurance have verified that the information they need is present in the drug prescription message. They have explained the advantages they expect from this project to the co-ordination group in the drug prescription project.

Today the health insurance pay in advance to the pharmacies. This implies that if the patient does not fetch the drug at the pharmacy, the health insurance have to pay anyway. This is no problem for the pharmacies. On the contrary, the pharmacy can then sell the drug to other patients. But the health insurance do not, of course, want to pay for more than they have to. To check whether a prescribed drug is actually delivered at the pharmacy or not is not possible to achieve by only considering the contents of the message itself; one has also to consider the system architecture.

One has to exploit the flexibility a standardised message always allows, to adjust the surrounding use -- in this case the system architecture. In the project, the health insurance have suggested a solution where all prescriptions are transmitted to a data base instead of directly to the pharmacies. When a patient arrives at a pharmacy to get her prescribed drug, the pharmacy finds the prescription in this data base. The whole idea is then that the health insurance should pay only when this transaction is actually made, not when the GP store the prescription in the data base.

The new computer systems and routines at the health insurance will also affect GPs, pharmacies and laboratories. The health insurance will now be capable of checking every re-payment they receive. They will also be capable of controlling whether the job is done properly or not, and they will also be capable of checking the patients as well. If the insurance find irregularities in the system, there are several ways this could be conveyed. The point here, then, is that the technical capabilities of the computer system of the health insurance go far beyond what is conceived as socially acceptable. To neglect the balancing act between what is technically possible and what is socially feasible is to bypass the real problem.

New and stricter instructions could be worked out, or the actors could get some kind of a report when unnecessary or too expensive tests are being done. One could in the case of laboratory orders imagine a kind of notification to the GP like "This analysis costs a lot of money and was unnecessary to do". But neither the GPs nor the suppliers of their systems would likely accept information that criticise the job done by them. They see this as intrusion into their professional responsibility. The problem is to avoid having too minute control mechanisms; the control needs to operate on aggregated information. A solution might be an accumulated evaluation of the work practice of the GP at a certain time according to how many unnecessary and also expensive tests done. It is most likely that this kind of evaluation would be done in some way when all this information is electronically available.

Conclusion

The challenge for actually achieving participation during the development of an II is both important and difficult. It is important because IIs are expected to have far-reaching economical repercussions. It is difficult for a number of reasons, some of which have been illustrated above. In summing up, we highlight two of these.

The design and deployment of an II will involve non-homogenous set of actors with distinct interests and perceptions of the II (Neumann and Star

1996). To function communicative, one nevertheless has to reach a consensus on the standard, a consensus which is problematic in the absence of a central authority. The identification of prescribed drug item illustrates this problem. This raises the issue of how to organise II development as a collective effort with reasonable influence from involved actors (Kahin and Abbate 1995).

Standardisation of II is not at all purely technical (Hanseth and Monteiro forthcoming). More specifically, the process of standardisation is aligned with a broader, international effort as the question about economic information illustrates. In terms of user participation, this implies that users are required to relate to how the design fits in geographical and application areas where they have little or no experience. The fact that the description of the standard is to oblique to users due to the technical jargon amplifies this problem.

The only viable strategy for developing IIs is, programmatically stated, to design and deploy them evolutionary (Hanseth, Monteiro and Hatling 1996). At every step, a sufficient amount of practical experience and learning-by-doing need to be acquired before moving on. The pressure for universal and "tidy" solutions need to be bridled for this to take place.

References

(antonelli92) C. Antonelli, editor. *The economics of information network*. North-Holland, 1992.

(baroudietal86) J. J Baroudi, M. H. Olson, and B. Ives. An empirical study of the impact of user involvement on system usage and information satisfaction. *Communications of the ACM*, 29(3):232-238, 1986.

(Carlsen 93) *Proposal to requirements specification for the GPs application* (In Norwegian), Christian Redisch Carlsen, 1993.

(CEC 93) *Growth, competitiveness, employment. The challenges and ways forward into the 21st century*. White paper, Commission of the European Communities, 1993.

(CEN 92a) Minutes and associated papers from *3rd CEN TC 251 WG3 Meeting* (14-16.02.92).

(CEN 92b) Minutes and associated papers from *5th CEN TC 251 WG3 Meeting*;

Appendix 5: Report from PT008 - Laboratory Messaging - revised set of documents (16-18.10.92) (CEN 93a) CEN/TC 251/WG 3/PT008/N93-185: *Messages for Exchange of Clinical Laboratory Information, First Working Document* 1993-10-11 (Red Cover Procedure).

(CEN 93b) CEN/TC 251/WG 3/PT008: EDIFACT *Message Implementation Guidelines* (October 1993).

(CEN 93c) European Standardisation Committee (CEN*)*: Directory of the European standardisation requirements for health care informatics and programme for the development of standards: CEN/TC 251 *Medical informatics* (version 1.7 June 1993).

(ciborra96) C. Ciborra, editor. *Groupware and teamwork*. John Wiley & Sons, 1996.

(Davidow 92) *The Virtual Corporation. Structuring and Revitalizing the Corporation for the 21st Century*, William H. Davidow and Michael S. Malone. Harper Business, 1992.

(grahametal95) I. Graham, G. Spinardi, R. Williams, and J. Webster. The dynamics of EDI standards development. *Technology Analysis & Strategic Management*, 7(1):3-20, 1995.

(Hanseth and Monteiro forthcoming) Inscribing behaviour in information infrastructure, O. Hanseth and E. Monteiro, *Accounting, Management and Information Technology*, To appear.

(Hanseth et al. 91) How Can Standardisation of Health Care Information Exchange Succeed? Ole Hanseth and Gisle Hannemyr. In *Proceedings from the 14th IRIS*, Dept. of Computer Science, University of Ume E5, Sweden, 1991.

(NAF-Data 94) Requirements specification for implementation of an EDI-module for drug prescription in the pharmacies application (In Norwegian), *NAF-Data*, (February 1994).

(KITH 91) *KITH*: Comments to the previous implementation guide for the laboratory report message, (Nov./ Dec. 1991).

(KITH 92) *KITH*: Report from the pre-project EDI-based exchange of drug prescription (In Norwegian), (Nov.1992).

(KITH 93a) Handbook for EDIFACT-based prescription messages, version 1.0 (In Norwegian), *KITH*, (June 1993).

(KITH 93b) Minutes from the 2nd Co-ordination-group Meeting (7.9.93) (In Norwegian), 1993.

(KITH 94) *KITH* Information model for laboratory communication - preliminary - Appendix 4: Comments to (CEN 93a): (February 1994).

(lehr92) W. Lehr. Standardisation: understanding the process. *J. of the American Society for information science*, 43(8):550-555, 1992.

(miller92) S. E. Miller. Political implications of participatory design. *In Participatory Design Conference* (PDC) '92, pages 93 - 100, 1992.

(mumford84) Enid Mumford. Participation - from Aristotle to today. In T.M2EA. Bemelmans, editor, *Beyond productivity: information systems development for organisational effectiveness*, pages 95-104. North-Holland, 1984.

(Schuler and Namioka 1993) *Participatory design~: principles and practices*, D. Schuler and A. Namioka (eds.), Lawrence Erlbaum Ltd., 1993.

(sclove92) Richard E. Sclove. The nuts and bolts of democracy: democratic theory and technological design. In L. Winner, editor, *Democracy in a technological society*, pages 139-157. Kluwer Academic Publ., 1992.

(Star and Ruhleder 1996) Susan Leigh Star and Karen Ruhleder. Steps towards an Ecology of Infrastructure. *Information Systems Research*, 1996.

(Statskonsult 92) *EDI transmittal of prescriptions*. Final report (In Norwegian), (February 1992).

(winner92) Langdon Winner, editor. *Democracy in a technological society*. Kluwer Academic Press, 1992.

11 From System Development to Organisation Development? - On the Role of User Participation in the System Development Process

EDELTRAUD EGGER
VIENNA TECHNICAL UNIVERSITY,
VIENNA, AUSTRIA

Abstract

There is a new term in modern software-engineering which becomes more and more important: User-Participation. The consideration of the end-user of a computer system during the development process underwent a serious change: from using the users as information source only for defining the functionality of the system towards an interactive process of combined organisational and system development. The following article will analyse the organisational conditions under which computer systems have to be developed. This leads to a critical assessment of the possibilities of user participation.

Introduction

The idea of user-participation can be placed in the sixties when the general software-crises lead to a more open discussion and more structuring of the developing process. Until that software was developed without considering the requirements of users except during the problem specification, where the users were important knowledge sources. The resulting products often were not adequate and therefore lead to only little acceptance of the users. This was the first step to the stronger consideration of users from the system designers' standpoint. On the side of firm-owners there was a financial reason for complaining about the inadequate products: They simply did not want to spend a lot of money in a software nobody wants to use.

Though in the meantime user participation seems to be an indispensable component of the software developing process, it is quite amazing that there are only little practical experiences and evaluations. This forces one to assume that the practical realisation induces a lot of problems. An interesting source of new approaches to the participation of users offer the Scandinavian countries, where many software-developers are already very familiar with methods and concrete realisation of it. The analysis of the results there will verify the theoretical organisational framework described in this paper determining very strongly the way, possibilities and limits of user-participation.

Therefore the first chapter will describe the main organisational issues, while in chapter 2 the internal conflicts of the participating users are evaluated. Finally chapter 3 comes up with concrete implications for system-designers.

The organisational framework for system development

Division of labour implies the establishment of hierarchies in firms: There are persons doing jobs and such ones controlling that the jobs are done. In other words there are workers and decision makers. This means that in all enterprises we find hierarchies as organisational structures.

At the different levels there are different degrees of autonomy and competencies. At the lowest level the workers execute tasks defined by persons at higher levels. Competence is limited to weak ad-hoc decisions if necessary. The next level is occupied by managers who have to control that work is done satisfyingly. Besides that, they have to assess measures to increase productivity. One way of increasing rationality and efficiency is the introduction of information and communication technologies. At the top stands the firm-owner with the decision power concerning all decisions of the enterprise. Microeconomic literature shows that in a company's reality working groups - doing what Galegher and Kraut (1990) called 'intellectual teamwork' - do not exist. Especially the statement on semi-autonomous groups being "only loosely coupled with the surrounding organisational environment" has to be doubted. Contrary to this the organisational frame-work of a firm will determine emphatically the working conditions of the various groups.

Generally it can be said that in most cases the decision if new information and communication technologies should be introduced at all is made by the firm's owner. So there is no user-participation. This idea can be limited practically to the introduction process, where the functional modules of the system are defined.

According to Schakel (1985) there are functional, psychological and economic reasons why the users of a computer system should participate in the design and development of software. Actually all these reasons can be summarised under the economic principle: Aim is to save cost.

As a matter of fact cost plays an important role in the software-engineering process: Tavolato and Vincena (1984) have proved empirically that 56% of all failures in system-development can be lead back to failures in problem analysing. The reparation takes 82% of the cost of the total error-handling. This makes clear that considering the end-users of a system will save a lot of money:

Software-engineering cannot be seen as single measure, but must be seen as a process with the aim to optimise the whole working-system. This implies that the introduction of information systems is strongly interwoven with organisational change (Galegher and Kraut, 1990) concerning work-content, rationalisation, labour intensity, changes in qualification - which will be discussed later. Pre-condition for such a change is concrete and detailed knowledge on problem specification and on the organisational framework. The end-users do have this knowledge. Therefore they can serve as experts and communication partners during the phase of problem specification and as control instance on different development-levels. This method leads to decreasing time and lower risks of wrong specification (Schakel, 1985). Duell and Frei (1986) suggest that co-operation in changing the own working conditions goes hand in hand with increasing qualification. This means that the introduction of new technologies is not seen as a coercive force but as possibility of co-determination which leads to higher acceptance which again implies lower counter-reaction (Tavolato and Vincena, 1984). The extent of user-participation is determined by the firm's owner, who has to make a cost-utility computation. Heilmann (1981) classifies the following degrees of participation as below:

No participation	No pre-information for concerned users. Pre-information of the concerned users.
User hearing	Opinion of users is heard and eventually considered.
Active co-decision	Users have veto right.
Active participation	Users participate in the decision and design process.
Autonomous design	Users design the system autonomously.

Evidently the degree of user-participation will depend very much on the internal organisational structure of a firm: rigid hierarchies will not induce a real co-determination, totally autonomous groups will rarely be found in firms.

Barley (1986) considers the introduction of information systems and the resulting discussion process as disturbance of the established organisational structure: competencies and responsibilities have to be re-negotiated. By defining the user-participation extent the scope of bargaining is determined too by the firm's owner. Since there are in general conflicting interests in a firm, there is also a conflict potential being hidden in the discussion on designing the information system. Contrary to Luhmann (1984) it is quite reasonable to assume that the result of such fighting is not the elimination of the underlying conflicts but some compromise. Since it is clear that moderating such a sensitive process requires special qualifications, it must be doubted that system designers are the right persons for this kind of job. Moreover it seems to be likely to have an experts' team,- which again is a question of money.

Consequences for system-designers

System designers are hired to design and develop computer systems. Coming from an area which in general does not have much in common with the application field, a major difficulty for them is to get familiar with the particular problems. Therefore a principal guideline for the software engineering process was developed, known as the classical "waterfall-model" (Figure 11.1):

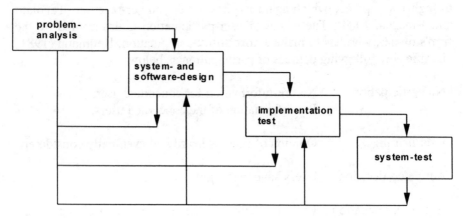

Figure 11.1   The waterfall-model

The software development process is seen as step by step procedure, where in principle each sequence of steps can be repeated itteratively till a satisfying solution is found - evidently the cost of iterations sets boundaries to the number of iterations.

User-participation is limited to the first phase, namely to the problem analysis. In order to define the functions of the computer systems the users have to serve as information source, describing the current working process and formulating wishes for improvements. It is the system designer who takes this information as starting point for designing a system-proposal. After checking it with the users (and eventually correcting the design) the implementation is done. A brief test phase concludes the development process,- the system is installed and from time to time maintained.

The major flaw of the waterfall-model is the neglect of the underlying organisational structure of a firm. The results are either exploding cost (due to numerous iterations) or inadequate solutions (which are not used as expected). Especially in the Scandinavian countries this deficiency has been realised and a remedy was developed: the so-called "prototyping" approach (Figure 11.2):

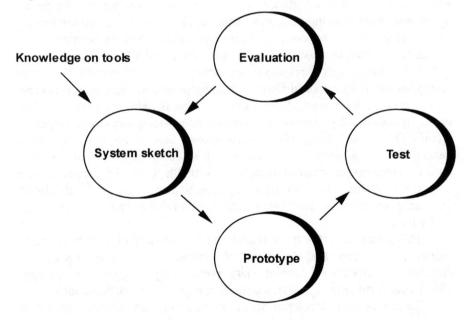

Figure 11.2   The prototyping approach

The basic idea of the prototyping approach is the permanent involvement of the future users of a system. Starting with a system sketch -

which is designed by the system designer (having the technical knowledge on available tools) together with the users, this sketch is implemented. After a testing phase with users (which is supervised by system-experts or tutors) the evaluation of the first prototype will lead to the extension of the system.

System development is very strongly understood as organisation development: The design and introduction of a computer system implicate changes in the organisational structure and vice versa. Therefore some methodological problems arise. The Work Research Institute in Oslo developed the method of a "search conference" (Bergman and Thoresen, 1988): People working in an organisation are meeting each other in order to discuss and design their working conditions. Working groups are built (organised either vertical, all members of a department, or horizontal - by function) which do have to work on certain tasks. The results are presented and discussed in plenary meetings. Strict rules are worked out which have to be followed. A research team is supervising the discussions and moderating it, but is not allowed to interfere as long as the rules are not broken.

This general action-oriented method can also be applied when developing a computer system. The various vertical and horizontal groups have to discuss the functionality of the system and they have to decide co-operatively how working conditions shall be changed. As can be seen easily the pre-condition for action-oriented user-participation is the permanent co-operation between the system designer (or teams) and all users.

Even though this approach is academically very interesting, the firm's reality seems to look a bit different: Binding people in discussion processes means keeping them away from work. But this is only one problem when trying to realise the presented method of user-participation. Some empirical studies (Duell and Frei, 1986) have shown that prototyping and user-participation in practice are confronted with a lot of problems: Its iterative nature makes project-management very difficult. Since the purchaser of a computer system wants to have a budget-calculation, it is absolutely necessary to estimate cost (also including working-time used for user-participation etc.).

This means that it is more realistic to see prototyping as the waterfall-model (concerning the definition of objectives and project-schedule) enriched by aspects of "evolutionary prototyping" (Egger and Hanappi, 1993), (concerning the necessary underpinnings of user-participation).

Seeing system development as organisation development means to meet the firm owner's and the users' requirements. Therefore a cost-benefit analysis has to be made in order to find solutions which satisfies both parties in the firm. This can happen only if there is room for negotiation and discussion on anticipated changes in the organisation (Heilmann, 1981).

Evidently a discussion of all the problems appearing in an "organisation-oriented" system development would exceed the scope of this paper. Nevertheless one important starting condition shall be dealt with: the decision which users will participate in the development process. In particular the following questions have to be put:

- Which users will participate in the design process and how are they elected?
- What rights do user have during the design and development process?
- To what extent are users involved in the design and development process?

Often it will not be possible to involve all concerned persons because they are too many. So one starting point will be the choice of users which can be made by defining criterion. The different methods to do so have different advantages and disadvantages:

All concerned users elect their representatives: The participating users obviously will elect a person of their trust. On the one side this is some sort of democratic control, on the other side this person will not necessarily be the most competent one (even though the users would probably consider this aspect when choosing a representative).

If the internal political structure consists of elected union-representatives they often also are automatically the representatives being involved in the design and development process. As practical experiences (Barley, 1986) show it can be that those persons do not have the technical qualification necessary.

Volunteers mostly do have the knowledge and the motivation, but it can be that the concerned users do not accept them as representatives because of lack of trust. Considering the enormous effects of introducing a technical system it is quite reasonable that the users must have the certainty that their representatives will not let them down in critical decision-situations. For similar reasons representatives chosen by the firm's owner will not be accepted.

All in all it can be said that it would be the best to involve all concerned users. If this is not possible because of the great number the most important point is that the users do have the feeling that their elected persons do represent their interests. This counts more than technical qualification which can be offered and acquired during the discussion and designing process.

Conclusion

As shown in this paper the development of a system is a far-reaching intervention into given organisational structures. Therefore it is necessary to face the internal conflict potentials and to pay attention to the conflict resolutions. Since a firm's owner is interested in investing in a profitable way, the development of a system has to meet some given financial conditions. As the experiences in software-engineering of the last twenty years show, it is also very urgent to consider the requirements of the future users in order to prevent counter-reactions.

The classical waterfall-model does neglect the user-participation side and the prototyping approach does not consider the economic framework of system-development. In order to close these deficiencies a new approach combining both methods has to be applied.

It is self-evident that this model demands high multi-disciplinary qualifications of system-designers and a lot of practical experiences. I am afraid that at the moment the system-designers' education and training do not take these abilities into account adequately, but I am sure that considering system-development as organisation-development will become an attractive and challenging view-point.

References

Galegher J. and Kraut R.E., Technology for Intellectual Teamwork: Perspectives on Research and Design, in: J. Galegher, R.E. Kraut, C. Egido (eds), *Intellectual Teamwork. Social and Technological Foundations of Co-operative Work*, Hillsdale, NJ, 1990: 1-20.

Schakel, Human Factors and Usability - Whence and Whiter? in: Bullinger (ed.), *Software-Ergonomie '85*, Mensch-Computer-Interaktion, Stuttgart, 1985.

Tavolato P. and Vincena, A Prototyping Methodology and its Tool, in: Budde et al. (eds.), *Approaches to Prototyping*, Berlin, 1984.

Bergman T. and Thoresen K., Can Networks Make an Organisation?, *Proc. CSCW-Conference*, Oregon, 1988.

Duell and Frei, *Leitfaden für qualifizierte Arbeitsgestaltung*, Köln, 1986.

Egger E. and Hanappi H., Competence-Switching Managed by Intelligent Systems, *Proc. of ISMIS'93* , Trondheim, 1993.

Heilmann, Modelle und Methoden der Benutzermitwirkung *in Mensch-Computer-Systemen*, Stuttgart, Wiesbaden, 1981.

Barley S., Technology as an Occasion for Structuring: Evidence from Observation on CT Scanners and Social Order of Radiology Departments, *Administrative Science Quarterly* 31, 1986.

Luhmann N., *Soziale Systeme*, Frankfurt a. Main, 1984.

12 CSCW in Information System Application Development

CHRIS DIXON,
STAFFORDSHIRE UNIVERSITY COMPUTER CENTRE,
STAFFORD, ENGLAND
TONY STOCKMAN
STAFFORDSHIRE UNIVERSITY COMPUTER CENTRE,
STAFFORD, ENGLAND

Abstract

Group working is a firmly established work practice, not least in the computing industry. Computer supported collaborative work, or CSCW, is a young discipline growing in importance. Surprisingly, there is little software support aimed specifically for use by work groups, which consist of developers and users who are working jointly on the production of new applications. In this paper, some of the issues surrounding CSCW are investigated. Arguments are put forward for features of a support tool that would be beneficial in this context. The practice of prototyping as a disciplined, managed part of the software development cycle is examined as a particular case where CSCW systems could be applied as a documentation, argumentation and assessment support tool.

An ORACLE-based experimental system which supports some of the above features, implemented to amplify the general treatment of the topic is described. The system provides explicit support for the different roles played by various members of the work group involved in the development of an application, and so provides documentation support for the roles and responsibilities of different work group members involved in the prototyping process. How this system can help in the management and control of the prototyping process is discussed, together with how the system may be used to maximise the benefits of prototyping without loosing control of the development process.

In addition to supporting the different roles of work group members, and providing prototype version control, the system includes specific support for the assessment of the usability of the application being developed. It allows for the selection and application of usability criteria with respect to each prototype developed, and so integrates and provides documentation for the processes of prototype production and evaluation.

Introduction

When man faces a problem too complex for him to solve alone, he traditionally looks to work in unison with others. From a team of hunters stalking a difficult prey, to a group of engineers designing a complex piece of machinery, there are examples throughout history of co-operation removing barriers which had stood in the way of the individual. This concept - of pooling resources towards the achievement of a common goal is firmly ensconced in today's working practices. Many benefits may be claimed from the collaborative approach. a group has superior problem-solving power to an individual working in isolation, and quick feedback from ideas and interaction with peers may reduce mistakes, relieve boredom and raise morale [Schage 1990]. The group has a range of perspectives on the problem, and will pick up on important issues that the individual may miss.

As anybody who has worked in a group will testify, the co-operative approach brings its own set of problems. Co-ordination of resources and communication between group members become significant factors, and may represent time and cost overheads. Determining the optimum size, composition and skill profile of a group is critical if the potential benefits are to be reaped, but is often given little attention.

The discipline of Computer Supported Collaborative Work, (CSCW), is dedicated to understanding the work patterns of people operating in groups, and to supporting such group work through the use of appropriate technology. Computer systems which do this are often referred to as Groupware. The terms Groupware and CSCW are subject to a variety of definitions, and are used interchangeably by many authors. Groupware systems support common tasks and shared environments in different ways, and to different degrees. Any attempt to measure the level of group support will necessarily involve consideration of several dimensions by which support may vary. The most common classification of groupware systems uses two such dimensions; time and location. Members of a group engaged in a common task may be located in the same office, or distributed around several sites. Also, the interaction between members of the group may be held in real time, such as face-to-face or by telephone, or asynchronously, for example via mail. These space/time distinctions are the basis for a widely accepted taxonomy of groupware systems. Examples of these classes of Groupware systems include: computerised meeting rooms which support synchronous co-located interaction (same time/same place), while video-conferencing facilities are for synchronous remote communication (same time/different place). The support offered by electronic mail is principally asynchronous and remote (different time/different place), while a shared database application is asynchronous, and may support co-located or remote users.

A comprehensive groupware tool might ideally support all four modes of group working. Many Groupware systems indeed cross the boundaries between two or more of the categories. Dix et al [1993] proposes an extension to the taxonomy in which the time axis is broken down into four rather than two divisions - 'concurrent synchronised', 'mixed', 'serial' and unsynchronised. There are certainly arguments to support the corresponding breakdown of the space axis to a finer degree of granularity. The time/space matrix is simplistic, but has been found to be useful, and is mentioned in most texts on the subject.

Groupware for software development support

It is perhaps curious that CSCW is such a recent area of study for software designers, given that the predominant work unit for software developers themselves has long been the co-operative group. Our own experiences of working in small groups in a number of commercial and academic environments, together with the documented evidence of case studies such as that of (Walz, Elam and Curtis 1993] who studied the machinations of a software design team over the course of four months while they were engaged in the specification and design of an object server system requested by in-house users, identify a number of points which consistently arise when examining the dynamics of group design. These points can be outlined as follows:

- When one exists, the prepared agenda for a meeting is often not followed. For example meetings scheduled to be administrative decision-making sessions very often can become technical discussions or the agenda can be in some other way hi-jacked by one or two attendees.
- Decisions made in one meeting are often forgotten by the next. Even when decisions are remembered, the reasoning that led to them is often forgotten, and the issues have to be re-discussed.
- Related to the above problem is the fact that some group members inevitably miss meetings, and are therefore unaware of some previous decisions.

- Meetings can frequently be dominated by members who are technically strong, even when non-technical issues are being discussed. This phenomenon is particularly serious when it leads to important user requirements being ignored or undermined, so reducing the value and acceptance level of the delivered system.
- The attendees of meetings are often inappropriate. Whether or not a particular individual attends a meeting should depend first on their likely overall contribution to the meeting achieving its objectives, and secondly, though often importantly, whether they are likely to benefit significantly personally from attending. Knowledge profiles may be useful in determining whether specific individuals should be involved in a particular series of meetings.

The above points can be summarised into two key problem areas as follows:

- Ideas and decisions arising from group project meetings can be hard to capture and document in such a way that the group can assimilate and progress them reliably.
- A lack of initial thought being given to the structure of the team can lead to awkward group dynamics, domination by one or more individuals and redundancy of effort.

Continuity between meetings and design rationale

The loss of ideas and decisions, problem A above, seems particularly telling. It could be argued that the keeping of minutes in the time-honoured manner would prevent this loss of information between meetings, but point number 1 above suggests why this often does not happen. Meetings in which the agenda is abandoned and complex technical issues are discussed in an unstructured way, are bound to be difficult to minute. In the early stages of an application's development, the need for collaboration is at its greatest because the task is only loosely defined and the risk of divergent ideas is high. At the same time, the co-ordination of the collaborative effort is at its most difficult, especially when there are no clearly defined roles and responsibilities within the team. Design teams have historically been expected to manage their collective 'memory' in an ad-hoc manner. Formal documentation (such as models, specifications, program listings and manuals), even if produced with the help of CASE tools, all represent outputs from the design process, and say little about the arguments of the process itself (why did we do it like that?), (what constraints were in place that caused us to take this series of options?) etc.

The discussion above illustrates that face to face meetings are not always an effective way of progressing the level of collective understanding in a group engaged in a complex task. This is interesting from a CSCW point of view, as the motivation behind many groupware tools is to achieve the communicative capacity of a face-to-face meeting when the real thing is not possible (because of time or location differences between group members).

Software development teams can therefore benefit from groupware tools that are designed to capture the rationale behind design decisions. Tools have been proposed [Krasner et al 1991 and 1992] which could take the intellectual processes of design as input (videotapes of meetings, conversation transcripts), and store this information in such a way that information can be retrieved in a convenient manner. These would fall into the category of argumentation tools. A further possibility is to strictly limit the use of meetings for discussion of technical issues. A groupware tool allowing team members to record issues, positions, problems and proposed solutions, which can then be inspected and progressed by other team members, could take the place of many group discussions. Such a tool would not need to offer support for same-time group work, although advanced examples of synchronous tools do exist. To make a contribution, in contrast to the usual practice in face-to-face meetings, the contributor would be compelled to formulate his or her thoughts, and would have the opportunity, free of interruption, to express his or her arguments and to view the ideas of others. The responsibility for translating thoughts and findings into decisions would have to be allocated, perhaps with different team members holding responsibility for different areas of development. This leads on to the second area of concern arising from the discussion above.

Composition and structure of a design group

The type of group work with which CSCW is concerned, and to which software design teams usually adhere, tends to preclude a rigid structure or a strict demarcation of responsibilities. The group is engaged in co-operating towards a common goal or task, and an overall understanding of the task is considered essential for each team member [Galegher et al 1990]. In large developments, the task may be divided into sub-tasks, each of which becomes the common goal of a sub-team [Schage 1990]. In reality, team members will take on certain roles, whether these are officially allocated or assumed by default. The dangers of leaving the role assignment process unmanaged are illustrated by the problems mentioned above. Meetings (many of which were never intended to be technical in nature) become dominated by those with the greatest technical knowledge, or at least those able to express their knowledge. Other participants may become bored or subdued. The upshot can be that the technical gurus are allowed to lead the decision-making process even in non-technical areas, and even in the domain of customer requirements! The roles of all other team members can become blurred, or close to nil. The composition of software design groups according to breadth of experience, skills, personality and other characteristics is an area studied in depth by Boehm [1981] and others [e.g. Galegher 1990]. Suffice to say here that it appears to be desirable to dedicate some thought to the roles that team members are best equipped to play or most likely to assume once the group starts to interact. This in no way means that people without previous knowledge may not be included; many organisations will wish to include inexperienced members to push them along the learning curve. If this is managed, and made known to the group, the negative notion of 'slack' in the team becomes a positive one of 'skill development'. A Groupware tool designed to assist such a group would usefully have the facility to define and support roles and responsibilities in a supportive and flexible manner. By this, we mean that roles and responsibilities should be user-definable, people should be allowed to assume different roles in different projects, and some form of access control to the tool should be implemented to help reduce user-to-user conflicts. Ellis, Gibbs and Rein [1991] note that access constraints are most usefully specified in terms of roles rather than individuals in this way, a certain amount of the required flexibility is inherent. Rather than having access privileges granted or revoked, the role(s) played by the individual in each project determine his or her rights.

Prototyping as a group activity

Prototyping is an iterative process which involves recursion through the cycle of 'user requirements-> design->code->user review' a number of times, preferably quickly. Good communication is of the essence, and this is hampered by the traditional environment with its departmental divides and conflicting requirements at the departmental and organisational levels. There is still widespread suspicion of prototyping among software developers, many believing that it is only appropriate for small systems, and that it cannot be incorporated into a conventional development life-cycle. Ian Graham [1991] argues for the value (his word is 'necessity') of prototyping from two angles. Firstly, that it produces improved human-computer interfaces, and secondly that it facilitates communication between developer and user with the result that specifications of requirements are fuller and more accurate. These two considerations are closely intertwined. The user has to accomplish certain tasks in association with the computer and any other tools; in this context, effective human-computer interaction is a vital part of the functional requirements of a system. The discipline of human-computer interaction (HCI) is rapidly growing in importance, as the productivity benefits of highly usable systems and satisfied, confident users are recognised by management. Any development group taking these factors seriously is bound to engage in prototyping in some form, as the iterative evaluation of interfaces by future users is an integral part of the HCI discipline.

We have seen that application design groups could benefit from argumentation tools which help them to document the process of design. This discussion applies particularly to design teams involved in prototyping. It is important that a record is kept of the process, along with the results of assessments and the reasons lying behind design decisions. Capturing design rationale in this way seeks to improve the continuity between successive meetings of developers and end-users. The availability of such documentation may save considerable time during the maintenance phase of that applications lifecycle. Having a clear description of the reasons why particular design decisions were taken, including the context of those decisions and the alternatives considered, provides a more sound basis for deciding how appropriate it may be to reuse design options in future systems. Additionally, the ability to examine the reasons for particular design choices provides valuable material in a case study approach to training.

A prototyping group is not just the designers who build the prototypes. The approach requires commitment from key users, who must be prepared to dedicate considerable amounts of time to arranging assessments, resolving business issues, and facilitating communication between designers and other users. Using the terminology of prototype management described in Graham [1991] we may define some role players to be fully involved in the iterations of prototype design and user review, while others roles may indicate that they are involved only with the evaluation and requirements specification stages, or from a higher level with the process as a whole. The precision with which it is desirable to specify the individual roles of developers and users will vary considerably depending on the size and culture within development teams. The overall point in the inclusion of support for the definition of roles within the system is to help to address problem B identified above: to support clear consideration and documentation of the roles of the members of design groups.

The system described in the next section goes some way to supporting these aims, and also addresses the issue of group composition, and group role definition.

A CSCW system to support group design

The system we have developed aims to amplify some of the ideas expressed above. It represents the core of a support tool for groups of people engaged in prototyping; as such it is a Groupware tool, supporting the management and evaluation of prototypes, argument and decision documentation. The system is implemented as an Oracle7 application, using SQL*forms (version 3) and SQL*menu (version 5) as the programming environment. As a database application, it offers asynchronous support only for groups whose members may be co-located or distributed at remote sites. The main features are as follows:

- Support for the definition of group roles and activities.
- A rule base determining the activities permitted for each role.
- Restricted access to certain functions depending on role.
- The facility to record the development and progress of prototypes within an application.
- Prototype assessment according to user-defined criteria.

Applications, prototypes and functions

Users can record any application for which prototypes are to be developed. If desired, functions of the application may be recorded. Any number of prototypes may be associated with an application, but each prototype is defined as belonging to a single application. If desired, the particular functions implemented by a prototype may be specified. Each instance of a prototype is identified by a label and a version number. A more recent version of the same prototype will have the same label and a higher version number. The description, objectives and other comments may be recorded against any prototype, and these may be different from one version to another. The due and actual dates for completion of each prototype may be recorded.

People and roles

The basic identification details of any person who may take a part in the prototyping process are recorded in the Person entity. The role that they play with respect to any particular application is defined in the Role entity. This has a mandatory relationship to a single Role Type; this entity represents a reference list of possible types of role. The list is user-definable. A person may take different types of role in different applications. When a person record is entered, it is possible to specify whether he or she is primarily technical or non-technical. Similarly, when defining role types, it is possible to state whether it is principally a technical role (e.g. Lead Developer) or a non-technical role (Lead User). When assigning a person to a role, the system is able to check whether, for example, a primarily non-technical member of staff is being assigned to a technical role.

Prototype assessments and assessment criteria

Each prototype may be subject to any number of assessments, in which the prototype is evaluated using certain assessment criteria. The criteria to be used are not fixed - they are selected from a reference list, which is itself user-defined. There are two types of criteria - usability and evaluation criteria. Usability criteria are those which may be assessed by observation or measurement of some metric; for example 'the number of failed attempts to complete a task'. Evaluation criteria represent subjective assessments from the user; for example 'Satisfaction rating for error recovery'.

Assessment activities

When arranging an assessment, the people who will perform Assessment Activities are defined to the system. The possible activities which may be allocated are defined in the entity Activity Type. Each Assessment Activity therefore holds a reference to a person, to an activity type and to the assessment concerned. Rules may be defined relating the type of role that a person plays in an application, and the types of activity they may perform. A role type may permit many activity types, and any activity type may be permitted to holders of a number of role types. This many-to-many relationship is resolved in the physical model into an intersection table, Allowable Activities. The purpose of this is to ensure that Assessment Activities are allocated to staff authorised to perform them.

System functions and interface

The pull-down style of menu is used throughout the system, both to call forms when the user starts up the system, and to perform functions from within the forms.

The SQL*menu application is present in all forms. This offers the user the option of activating the menu and choosing functions from the pull-down menus as an alternative to using function keys to operate the forms. Menu operation is slower than use of function keys, but offers greater reassurance to the user. All pages of all forms in the system are defined as pop-up pages, with the view of the page anchored between lines 3 and 22 of the screen. This ensures that the menu options (even when not active), the title of the form and the message line are always visible to the user. The main menu offers five options:

1) Maintain Reference Data.
2) Maintain Personal Details.
3) Define Applications and Roles.
4) Define and Assess Prototypes.
5) Exit.

Option 5 is self explanatory. Option 2 calls the form 'Maintain Personal Details' (MTPERS). Option 3 calls the form 'Define Applications and Roles' (MTAPPS). Options 1 and 4 call the next level of sub-menu, also in the pull-down style. Option 1 offers the choice of calling the forms 'Maintain Role Types' (MTROTY), 'Maintain Activity Types' (MTACTY) or 'Maintain Criterion Types' (MTCRTY). Option 4 offers the choice of calling 'Define Prototypes' (MTPROTS) or 'Arrange Prototype Assessments' (ARRAST).

Discussion

To facilitate rapid application development, the system provides a means for an organisation using prototyping to record the progress of prototypes in an application development; to define the roles and responsibilities of those involved, and to assess the success or otherwise of successive versions of the prototypes against user-definable criteria. Using the system would represent a certain administrative overhead: reference data and personal details need to be maintained, prototyping plans and objectives described and assessment sessions arranged and recorded. However, any organisation undertaking prototyping in a controlled manner would need to record all these details somewhere. Given an appropriate working environment (for example, if all group members are able to access the shared database from networked PCs which also support most other aspects of their work), it would be feasible for the system to be the primary, or only, means of recording the prototyping data. In this case, the administrative overhead is minimal. The data model has been designed so that much of the legwork of running the system can be removed. Reference tables of role types, activity types, people and criterion types would need updating rarely. These, along with rules governing the allocation of activities, mean that much of the data entry involved in the day-to-day running of the system can be selected from restricted lists (e.g. using the 'list-of-values' function in SQL*Forms). This reduces the workload in terms of both thought and typing effort. features of organisations likely to benefit from such software support for group design include: an active quality program - an emphasis on proper documentation of progress, and evaluation of design deliverables and decisions, high usage of staff in work groups requiring careful attention to the structure and composition of the groups and the responsibilities of their members. The above would tend to suggest a large organisation, although smaller firms engaged in extensive information systems development may well fit the description.

As a more general evaluation tool the assessment part of the system is presented primarily for the support of prototype evaluation. There is no reason why other software should not be evaluated along the same lines. Different criteria may be used for assessment of such software, and this is supported by the system. As an example, most companies engage in evaluation of software packages before making a purchase. The method of evaluation, however, may be somewhat random, and is likely to be different depending on who has been asked to perform the evaluation. Use of this system would impose a structure on the evaluation process. Group members not directly involved in the evaluation could still play a part, by helping to select the assessment criteria, acting as guinea-pigs in assessments and commenting on the results obtained. Other software, such as existing systems, successful or otherwise, could be evaluated in the same way. There are clear benefits of a coherent approach to all software evaluation. Off-the-shelf software, and bespoke systems, old and new, are often run alongside each other, nowadays usually on the same screen. A consistent approach to 'look and feel' features between all types of system would help to smooth the edges of the user environment, and may well have productivity benefits. A firm interested in developing organisation-wide software standards could gain considerably from a systematic approach to software evaluation.

Conclusions

In this paper, we have outlined some of the general issues surrounding CSCW, and then concentrated in particular on systems to support software design teams. The arguments developed are then applied to groups engaged in prototyping as part of a design process. One conclusion from this is certainly that such groups could benefit from the appropriate use of CSCW tools. We have used a case study, and some personal experiences of software design groups, to put forward a case for a particular type of tool. A prototype system has been described which includes a core set of facilities to support the process of rapid application development. The tool described attempts to improve the cohesion and communications within design groups, provide some structure to the management of the prototyping process and capture design rationale to support future maintenance, training and reuse.

References

Boehm B. *Software Engineering Economics*. Prentice Hall (1981).

Dix A. et al *Human-Computer Interaction*. Prentice Hall (1993).

Ellis C. et al Groupware - some issues and Experiences, in *Communications of the ACM*, Vol 34, No 1 (Jan 1991) pp 39-58.

Galegher J. et al *Intellectual teamwork: social and technological foundations of co-operative work*. Erlbaum, NJ (1990).

Graham I. *Object Oriented Methods*. Adison-Wesley (1991).

Greif I. (ed.) *Computer supported co-operative work : a book of readings*. Morgan-Kaufman (1988).

Hartson H. & Boehm-Davies D. User interface development processes and methodologies, *Behaviour and Information, Technology* (1993) Vol 12, No 2, pp 98-114.

Jessup L. & Valacich J. *Group Support Systems : new perspectives*. Macmillan (1993).

Krasner H. et al Groupware research and technology issues with application to software process management, in *IEEE Trans on System Management and Cybernetics* (1991) Vol 21, No 4, pp 704-712.

Krasner H. et al Lessons learned from a software process modelling system, in *Communications of the ACM* (Sept 1992) Vol 35, No 9, pp 91-100.

Olson J. et al Computer Supported Co-operative Work : Research issues for the 90s, in *Behaviour and Information Technology* Vol 12, No 2, (1993) pp 115-129.

Schage M. *Shared Minds*. Random House (1990).

13 Enabling Group Work - Facilitation and CSCW Tightly Knitted

JOHANNES GÄRTNER AND ANDREA BIRBAUMER
VIENNA TECHNICAL UNIVERSITY,
VIENNA, AUSTRIA

Abstract

Facilitation and CSCW are two powerful approaches to improve group performance, participation, and interaction especially for non-routinised work. Still, little effort has been spent on their integration that is not far beyond computer support for standard facilitation techniques.

We thoroughly discuss decision making in work groups, and the limitations of each approach in supporting this. Based on our encouraging experiences with facilitated computer-supported group design in the strictly non-routinised field of shift-scheduling we show that integration goes beyond small improvements of each approach. Task adequate knitting enables efficient work in groups where it has not been possible so far.

Introduction

Group facilitation and CSCW have many features in common. The most obvious is that both have no precise definition and are best described by their aims and approaches.

Group facilitation aims to support groups. Facilitators intent to give support in different dimensions of group output ranging from primarily performance oriented support (e.g., facilitation in Joint Application Design [Andrew and Leventhal 1993; Wood and Silver 1989]) in a continuum to primarily personal outcome oriented approaches (e.g., helping participants to learn [Heron 89]). Facilitated group work is normally based on a formal appointment and voluntary acceptance of a facilitator and his/her role by the group members. Besides their skills and adequate behaviour facilitators use group-techniques (e.g., group-building, meta-plan) and their central role within the group process to accomplish their tasks. Facilitated group processes mostly take place in one or few numbers of meetings.

CSCW's special interest is "an endeavour to understand the nature and requirements of co-operative work with the objective of designing computer-based technologies for co-operative work settings" (Bannon and Schmidt 1991). This includes a broad range of more or less routinised work settings with much attention to problems arising by Cupertino within physical and within virtual organisational settings. It also includes computer support for task groups working at the same place at the same time.

Besides their common focus on group support both approaches consider the organisational context of group work to be crucial, namely the social, the organisational and the technical environment (e.g., Rabenstein et al 1990; COMIC 1993).

Still, little effort has been spent on their integration. Viller (1991) made a strong case on this neglection and explored some possibilities of supporting the facilitator in his/her work. The lack of integration and the necessity of research in this field is also criticised by Bostrom et al (1992) as well as by Vogel (1994). Today the facilitator's role within computer supported work group settings is mostly perceived as applicant of several standard group-techniques (e.g., generate, organise, evaluate), as a manager of problems caused by the group process and partially as a technical expert of the computer system in use. Computer support for facilitated groups is mostly oriented towards support for standard techniques (e.g., brainstorming), eventually extended by small features (e.g., Viller 1991: monitoring of participants' activity).

We argue, that the integration of these two approaches should and can go further. In spite of just supporting the facilitator in his/her role or just using facilitation as an addition to "normal" computer support for groups we propose a stronger knitting for complex group tasks. Such a knitting should allow efficient group work for tasks where actually it has not been possible to work in groups so far.

High complexity of group tasks may arise by several reasons. Firstly, by high technical complexity, i.e., if the generation, the evaluation etc. of possible solutions can't be done straight forward. Secondly, if things get difficult on the group and member level, i.e., when requirements are fuzzy, political conflicts arise and hidden agendas have to be dealt with (e.g., in the field of time management Egger and Wagner 1992). Both conditions are given in many organisational settings and important reasons for establishing groups to handle them.

Shift-scheduling is extremely complex in both dimensions. It is technically highly complex as solutions have to be constructed in an extremely huge solution space involving a high number of fuzzy, partially conflicting requirements. At the same time shift scheduling is deeply meshed with all kinds of interests, politics, and conflicts, that indirectly concern many important organisational aspects (Gärtner 1992). Currently, it is done by single experts that develops a proposal, which takes from a few hours up to several weeks. Afterwards that proposal is discussed in groups of people concerned. This goes back and forth until a solution is accepted.

In the next chapter we prepare a tight integration of facilitation and of computer support for group work by discussing groups and complex group tasks in their organisational context. Then we describe the focus and the limitations of facilitation and of CSCW-systems if used each on its own, and how they could be knitted together to allow facilitated group work, where it has not been possible so far. After a short introduction into the problem field of shift scheduling and our computer system, we finally describe our encouraging experiences in lab and field situations with this task adequate knitting.

This integrative knitting of facilitation and computer support goes further than additive usage of these existing approaches. It enables new effective ways of group work for complex tasks.

Groups and complex tasks

Before starting to discuss foci and limitations of facilitation and of computer support for groups it pays off to have a closer look on the underlying problem field.

Groups and decision making in organisations

We won't add another definition to the already overwhelming number of "group" definitions (e.g., compare Fisher, 1981), but just concentrate on groups, that work together to accomplish a complex common task. Such task groups are well established in all kinds of organisations to tackle complex, non-routinised tasks (e.g., design, organisational change). They are somehow well defined with respect to group-membership and their common task that includes a lot of decision making.

March (1991) building up on earlier work draws our attention towards problems of decision theory. Standard theories of choice view decision making as intentional, consequential action based on knowledge of alternatives, knowledge of consequences, consistent preferences ordering and a decision rule. He points out that some of the problematic assumptions of this theory have attracted attention (e.g., limits on the alternatives considered and the amount and accuracy of information that is available), but others even more problematic (e.g., the uncertain future preferences for the consequences of current actions) have been less considered. He draws the following implications for information engineering:

> First, a notable feature of investments in information and information sources that we observe in organisations is the extent to which they deviate from conventional canons of information management.

> Second, a major feature of organisational adaptiveness is maintaining a proper balance between the exploration of new ideas and the exploitation of old ones.

> Third, the management of life and organisations is probably as much a matter of managing ambiguity and interpretations as it is a matter of managing choices.

These findings of March are supported by Olson's et al. (1992) analysis of small group design meetings. They studied 10 design meetings of four projects in two organisations with differing numbers of issues involved. To their surprise in all meetings only 40% of the time was spent in direct discussions of design. Another 30% was devoted to taking stock of the progress trough walkthroughs and summaries and another 20% for pure co-ordination. The clarification of ideas - cross cutting the other classifications - took one third of their time.

The consequences of March's reflections on decision making in organisations are far reaching. Support for groups has to include elements of supporting choice in the view of choice of standard theory and at the same time (!) supporting management of deviations from this approach, management of idea exploration and exploitation, and the management of ambiguity and interpretations. Especially this holds for task groups for highly complex, non-routinised tasks.

Decision making in groups

March's analysis of decision making in an organisational context has to be further extended to decision making in task groups. The view of decision making is primarily oriented towards the results of decisions and the way they are reached. When discussing decision making in groups additional dimensions have to be considered.

McGrath and Hollingshead (1994) discuss the interaction of group and technology. Based on their earlier developed TIP theory (Time, Interaction, and Performance) they distinguish three interpretational dimensions of group functions:

- Groups as information-processing systems;
- Groups as consensus generating and conflict-resolving systems;
- Groups as vehicles for motivating and regulating member behaviour.

The dimension of information-processing pertains in their analysis primarily to the production-function of groups. The second dimension is the group-well-being function, and the third to the member-support function. They claim that groups are continuously and simultaneously engaged in these three dimensions.

These three dimensions enlarge the perspective on what has to be done in order to support groups in their work setting. Besides production oriented aspects of the work process, also group well-being and member-support have to be considered.

Group process

To make things even more complex, McGrath and Hollingshead (1994) add further aspects to be considered. They distinguish among four different working modes of how groups carry out the functions noted above:

- Mode I: Inception of a project (goal choice);
- Mode II: Solution of technical issues (means choice);
- Mode III: Resolution of conflict (i.e., of political rather tan technical issues); and
- Mode IV: Execution of the performance requirements of the project.

These modes should not be considered as a fixed sequence of phases but rather as potential forms of activity of groups. Flexible switching between these modes is part of the group process and has therefore to be supported or at least to be allowed.

Computer support and facilitation

The view of group work in an organisational context we developed above has led us far away from the classical view of decision making. In efficient group work

- A proper balance between exploration of new ideas and the exploitation of old ones has to be found.
- The management of ambiguity and interpretations has to be supported as well as the management of choices.
- Not only the production function, but also group well-being and member support have to be maintained.
- Different, flexible switching of modes of group work has to be supported.

In the following we discuss the consequences of this broader view of decision making and group work for technical and facilitative support as well as their integration.

Technical support

A number of different classifications of computer support for groups were proposed (e.g., DeSanctis and Gallupe 1987). We follow the categorisation used in McGrath and Hollingshead (1994) to distinguish between:

- GCSS: Group's internal communication support systems,
- GXSS: Group's external communication support systems,
- GISS: Group's information base support systems,
- GPSS: Group's performance processes support systems.

Research in the GCSS area is mostly directed towards spatial and/or temporal distributed group work. Little work has been spent on the support of face to face meetings. GXSS systems may play a role in face to face meetings (e.g., to get additional information quickly) but are not directly designed for this use.

GISS systems aim to improve the group's information base by supporting sharing and forwarding of information, that one member of the group already has. By this the group's potential information base is the union of the individual information bases. Two critical issues come up with this. Firstly, problems of information overload have to be solved. Secondly, under which conditions is information shared or not shared which is especially relevant in the broader view of group work and decision making (e.g., ambiguity). Actually, the main support coming up from such systems is directed towards the classical view decision making.

GPSS are directed towards group's performance processes. An example for such a system is SAMM (compare Dickson et al. 1992), which has an impressive number of functions to support group performance. It is designed to empower the team to create its own structure of work from a set of tools and techniques. The main focus is on group performance but the system tries not to get in conflict with other dimensions of group work (group well-being and member-support) and does not impose too many restrictions to different modes of usage. Still, these other dimensions and the switching and adapting to different modes are not supported.

To extend GPSS to give technical support in these later dimensions and modes seems problematic anyhow in the light of Dicksons et al. work (1993). They compared group decision support systems with respect to chauffeur driven and facilitator driven support. Chauffeur driven work was found to be more efficient, but they hypothesised that the reason for this was the style of facilitation and not in facilitation on itself. They concluded that to be effective in the judgement tasks considered facilitation had to be open and adaptive rather than restrictive. This causes more fundamental problems as it has to be considered technically most complex (if ever reached at all) to give open and adaptive facilitation support by technical means.

One further feature of the systems discussed is of interest. Even though these systems mainly focus on the production function of groups, they concentrate on very general features and techniques of group decision making. This may work fine when decisions can be made rather straight forward after the criteria's etc. have been fixed. But a relevant part of group work is not that simple. It includes the development of rather complex and - at least in many cases - domain specific alternatives and their evaluation.

Summarising, existing computer support concentrates on the support of simple decision making with respect to the production function of a group being open to flexible use (with respect to the other dimensions of group work and different modes of the group process) at best but without support in these aspects.

Facilitation support

The number of facilitation techniques and approaches is overwhelming (e.g., Rabenstein et al. 1990 that consists of 5 books of different techniques). Basically they concentrate on techniques for different modes of the group process (inception, goal setting etc.) and the dimensions of group work (production, group-wellbeing, member-support).

Much attention is spent towards open and adaptive facilitation. Besides the production function also group-wellbeing and member-support are considered to be important. Sentences like ("interrupts in the group process have to be handled first") are a good illustration of this. Much attention is also spent on the preparation of group work, i.e., who is participating, what are the aims, the setting, the timeline etc.

The high number of different methods and techniques to improve group work can be summarised into three main fields:

- focusation of group work.
- Visualisation.
- structuring of group work.

Focusation aims towards common work, i.e. it should be clear to all group members at any time what is on the agenda, who takes the floor etc., and towards what the contributions of the participants aim at.

Visualisation relies on a set of techniques that support focusation but also help to build up a storage of contributions and to support orientation within the group process (what are the topics of the agenda, the aspects of our problem, etc.). The used techniques allow a broad range of rather flexible modalities of visualisation that can't be met by existing computer technologies in speed and flexibility (e.g., drawing, using the whole room for visualisation).

The structuring of group work aims at a successful flow of the group process in its dimensions and in its modes. Much attention is spent towards this, especially in the dimension of time that is relevant in two aspects. Firstly, how much time should be devoted to which part of the work process. Secondly but even more important, is a good timing of each step (when and how long) in order to not disturb the group process. (e.g., a facilitator must not work for 15 minutes without talking to the group when everybody is waiting for results.)

Summarising, facilitation offers extremely flexible, open and rather adaptive approaches, tools, and techniques that can't be met in many features by existing computer systems. However, limits arise when problems and solutions get technically very complex, i.e. when the generation and evaluation of alternatives, proposals etc. is very complex and the focus lies on developing documents and plans and not on making decisions (e.g., in systems design, in planning). Then timing of group work does not work with the time needed for these tasks (e.g., to write a complete proposal and then discuss it again), which makes work in facilitated groups impossible.

Group work where it has not been possible so far

The weakness of facilitation applied to technically complex tasks does not allow group work in this field now or limits its scope strongly. This opens a new field of computer support for group work. i.e., the technical support should aim towards the reduction of technical complexity to make group work feasible in a field that could not be tackled by groups so far. This is a totally different approach from the existing support for face to face group work that mainly focuses on the automation of standard facilitation techniques and on the production function for (technically and organisationally) simple tasks. The reduction of technical complexity is a traditional aim of computer sciences. But this time it has to be done in a way that works with facilitation. The computer system has to give support (domain specific and general) in a way that does not interfere with the facilitation approach (e.g., visualisation, structuring). Both approaches have to be tightly knitted together.

Design for facilitated group work

In order to enable group work for complex tasks the computer support given must not interfere with the facilitation and must allow tight knitting of both approaches. The first thing that comes up is focusation and visualisation.

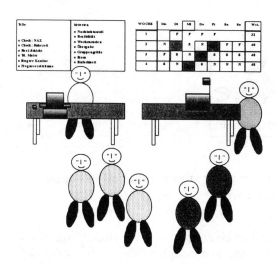

Figure 13.1   The setting of design process

Focusation and visualisation can be supported by computer. The alternatives discussed, analysis data, etc., have to be displayed by a projection system (e.g., in the simplest case by an overhead projector and an LCD display (Figure 13.1)) that every participant can follow and influence the planning process.

The system has to be designed in a way that the projection gives a meaningful overview of the actual topics concerned, which is part of the general visualisation problem (compare e.g., Gershon 1994), but has to be specifically done for the group setting and the task. It is not just a problem of visualising results but the visualise intermediate results within the work flow that includes topics like 'What could be done in the next step? What are our alternatives in the actual state?'.

The space given by current projection systems is rather limited. Therefore the number of items visualised by the computer system is restricted and it must be possible to choose between adequate presentation forms. This also constrains visualisation of other elements of the group work than those relevant to the production function. Such elements should be covered by classical visualisation techniques if otherwise overview or flexibility would get lost.

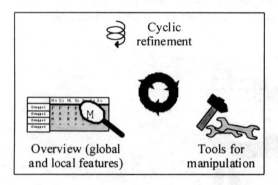

Figure 13.2   Basic elements of computer support that allows open structuring of group work

The second important question to be answered is the structuring of the group process. In order to allow open and adaptive structuring of group work the design approach for computer support is based on three elements (Figure 13.2), namely a cyclic design and refinements, tools for the manipulation of models and tools to support overviewing global and local features. The basic idea is to provide users a powerful workbench of tools with as little restrictions to the design and to the structuring of the design task as possible.

The critical issue in designing these elements is that they must not interfere with the group process in general and especially with the timing of the group process. Therefore it has to be transparent for the group members not to disturb focusation and not to involve elements of usage that are too complex with respect to the skills of the group members. Furthermore, it has to be fast and must not interfere with visualisation, and with structuring.

The case study: shift scheduling

We illustrate how group work may be enabled in a field where it has not been possible so far by tightly knitting facilitation and computer support that suffices the design requirements discussed above. Our application field is shift-scheduling which is technically and organisationally very complex.

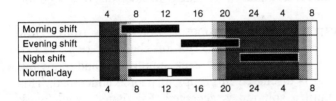

Figure 13.3   Shifts (black...working hours).

DAY	1	2	3	4	5	6	7	8	9	10
Group 1	M	M	E	E	N	N				
Group 2			M	M	E	E	N	N		
Group 3					M	M	E	E	N	N
Group 4	N	N					M	M	E	E
Group 5	E	E	N	N					M	M

Figure 13.4   Roster with the length 10 days (M ... Morning shift, E...Evening shift, etc.)

Shift Scheduling and its Organisational Context

Shift models (Figure 13.3, 13.4) are primarily defined by shifts and rosters. Shifts define duties (e.g., morning shift from 6 a.m. to 2 p.m.). There may be different shifts for weekends, or for certain times of the year and a more complex internal structure (breaks, stand-by duty, etc.). Rosters define which (group of) employee(s) has to work at which shift. The example in 13.4 is a roster with a length of 50 days (Group 1 starts with the first row switching to the second row after 10 days, etc.). Rosters with a more complex shift-sequence and longer duration (e.g., 48 weeks) are broadly spread.

The development of a "good" roster is a difficult but important task. Different shift-models have different effects on working conditions (health, income, social habits, etc.) of concerned employees and the company (e.g., costs, timelines). They also strongly influence work practices (meetings, information flow, possibilities for Cupertino, demands for co-ordination, etc.) (Gärtner 1992). The amount of shift work is high and in some branches still growing (Wolf and Vollmann 1992).

Scheduling is rather difficult and cost intensive. The number of requirements and constraints is very high (e.g., requirements by law, by company, by concerned groups), fuzzy and partially conflicting. The same holds for the mathematical complexity (a huge, sparse, discrete solution space).

Currently, one expert develops a proposal, which takes from a few hours up to several weeks. It involves a lot of checking (e.g., laws), counting (e.g., average working time of each employee) and drawing. Afterwards that proposal is discussed in groups of people concerned. This goes back and forth until a solution is accepted. Facilitated group work would be good to improve speed and quality of decision making but is not possible due to the time needed for the development of each proposal. Simple changes would make the group wait for hours.

Existing computer systems for this field concentrate on the handling of (few) explicit requirements and try to find optimal solutions (with respect to an utility function, e.g., Schönfelder 1992) with fully or highly automated systems (e.g., Nachreiner et al. 1993, Knauth 1987). They hardly can be used by groups. Their main disadvantages for group work are that their proposals are difficult to tune and that the computation time needed may go up to hours, and that visualisation is not supported very well.

The System: ShiftPlanAssistent - SPA

It is highly nontrivial to keep the overview of features of shift models (e.g., for 12 groups for 12 weeks). Designing by hand makes excessive counting and checking necessary, different representations are useful for different tasks and have to be written by hand currently. SPA supports overviewing in several ways that allows quick and easy feedback and supports refinement:

- Different representations (e.g., sorted by groups, by shift types, etc).
- Checking whether laws are fulfilled.
- Counting (e.g., number of night shifts, free Sundays, weekends; average working time; differences between working hours and operating hours).
- Graphical representation of features (e.g., night work per group).
- Computing possible cycles for specific features (e.g. how long would the roster be, if each group would have the same number of free Sundays?).

To provide users a powerful workbench with as little restrictions to the design as possible they may use a number of tools. These tools can be divided in tools for:

- direct manipulation (e.g., fill in shifts, groups etc.; copy, paste...)
- indirect manipulation (e.g., fill in shifts in the following way; give free time in the following way).

The latter one is a very powerful tool for quick and dirty design of a schedule, the first one is needed for tuning. Basic model features (e.g., number of groups of shift workers, operating hours) can be changed at any time. A screen shot of a simple session (in German language) is given in Figure 13.5.

The screen shot shows icons for refinement of basic features of the model at the very left (e.g., definition of shifts). In the middle there is a simple plan. Differences to operating hours and average working hours of the groups of employees are shown. (As the schedule fits well all the differences are zero.) On the right there is a toolbar and a menu for selecting shift-groups.

A number of tools (e.g., checking labor laws) are accessible via the menus. Even though the tools are very powerful most of the computational steps are finished within a few seconds not disturbing group work. Short-cuts are possible but generally not used as they would destroy transparency of use.

Knitting facilitation and computer support together

The planning process is done interactively in the group, supported by a facilitator and a driver of the supporting system. Other facilitation tools are used to co-ordinate work within the group (e.g., To do lists, prioritised requirements). Depending on the complexity of the planning problem design is done within a few hours or a small number of meetings.

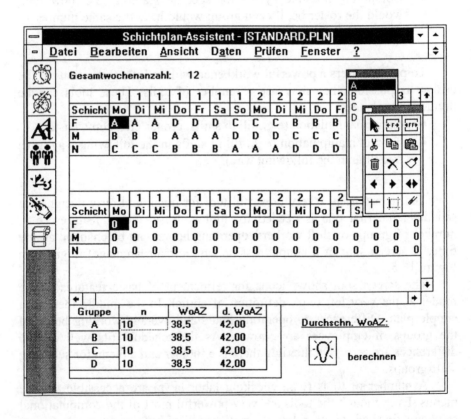

Figure 13.5   Screen-shot of the workbench of the ShiftPlanAssistent (SPA)

The facilitated design process concentrates on cyclic refinements of prototypes of shift-models and "prototypes" of requirement lists. This reflects the problem that designing shift models is both: a reflexive learning process on the actual prototype and a reflexive learning process on requirements. Prototypes of the model are discussed with respect to the requirements and are refined or rebuilt. Requirements are discussed regularly in the group to refine them.

As the facilitator's job in the case of a shift-planning process is rather complex (technically and organisationally), we want to name his/her main tasks. First of all his/her work has to be oriented simultaneously on

- the computer based planning process, the aims, the results wanted, etc.
- the group and working process.

To guarantee the flexibility of the facilitator to work on the planning process one moment and to switch to the group process the other moment, the facilitator (and if possible a driver) must have knowledge in both, the technical and the group working realm.

Let us imagine, e.g., a situation with several possibilities of further refinement of the shift-schedule. The facilitator notices, that the group situation is open, the members are willing for alternatives, but they do not know them. So he/she acts using his/her knowledge about shift-planning at the same time as he/she notices the status and the possibilities of the group. He/she has to make alternatives clear if the group members lack the skill to do this. The evaluation of these alternatives with respect to the organisational setting is done by the group members again. This technical domain knowledge is crucial in the field of shift-scheduling as it is technically that much complex and highly non-routinised to allow the participants to see which organisational choices are possible.

So the facilitator has not only to use techniques, his/her technical knowledge for presentations, etc. but his/her competencies concerning shift-planning and group processes. It is his/her task to see, what is going on in the group, in order to know, if e.g., a decision has to be made or not. It is his/her task to see, if it is the right time to go into a decision process. It is his/her task to see, which aims could be reached now and which aims have to wait. It is his/her task to time the steps of group work in an adequate way. He/she has to establish priorities during the whole planning process, also in the sense, whether to work on the factual level or on the level of relations. At the same time the actual state of the development of proposals, the technical alternatives given, and the tools that may be involved must be transparent for all group members involved.

The job of facilitation, as we see it, is demanding. On the one hand he should be able to see conflicts occurring and to manage them in the right way without suppressing them, on the other hand the facilitator should not try to make a therapy out of his/her job and additionally he/she has to guide the participants through technical questions. Anyway the facilitator has to "create" a productive working atmosphere with a satisfying result at the end. Members of such a group should be:

- Employees directly affected by the shift-model and staff on stand-by duty (either all or some representatives).
- personnel managers, shop stewards.
- department managers.
- foremen responsible for detailed assignment of work.
- a facilitator and a technician who "drives" the supporting system.

The size of the group may vary depending on the design tasks. i.e., sometimes the group may split to solve specific technical problems or to develop basic alternatives. Finally (and if necessary, for crucial decisions) all persons concerned (e.g., representatives of indirectly concerned departments) should be involved in the final checks.

Every member of the group should know, "who" the others are, they should be able to distinguish precisely between the functions and the interests being represented in the team. In this context all the members have to keep in mind the possibly different power and interests of their colleagues in the decision process as well. As we do not take for granted that the aims or even t h e aim must be clear at the beginning of the group process, we should only demand of the members to have their individual aim(s) reasonable clear and that they are willing to work on a solution.

Experiences

We used the approach described above in a number of meetings, including a lot of real applications, four "quasi-real" meetings with shift-planning experts, and four experimental meetings with shift-plan novices.

The main aim was reached. It was possible to do shift-scheduling efficiently within groups which had not been possible before. Practical relevant schedules could be developed in a few hours. Especially the overview of the schedule, the easy way of changing minor features and the checking and drawing features were considered to be very useful.

We had to overcome substantial problems to ensure focus especially for novices of computer systems or novices of shift-scheduling. In the first meetings we had not spent enough time to ensure enough overview of the basic elements of the system. Still the members were able to participate after some time. Adequate training at the beginning has to be considered as critical.

There is little experience in the given literature on the facilitation of task groups interacting with technology in the way we did it, especially for groups with highly conflicting interests. The same holds for the design of the computer system, where we are satisfied with the level reached but still found a lot of things to improve (e.g., handling of a high number of intermediate results). We are working on a great deal of improvements at the moment.

Conclusion

Computer support for groups - if designed adequately - may allow to work in facilitated groups where it had not been possible so far. The adequate knitting of these approaches brings up new requirements for each approach. Computer support has to be designed not to interfere with facilitation. This has consequences for visualisation as well as for the design of the tools to be used. They have to support overview and to allow fast, cyclic refinements. Facilitators have to develop their technical and eventually their domain specific skills and have to knit these approaches together tightly.

Further application fields of the knitting of facilitation and computer support to allow group work are (participatory) systems design, the planning of complex projects or the development of complex proposals whenever facilitated face to face group work seems adequate due the political, technical or organisational complexity of a topic. The benefits of our successful knitting suggest that it is worth the work.

References

Andrew D. C., Leventhal N. S. (1993) FUSION - Integrating IE, CASE and JAD: A *Handbook of Reengineering the Systems Organisation*, Yourdon Press.

Bostrom R. P., Watson R. T., VanOver D. (1992) The Computer-Augmented Teamwork Project, in Bostrom R. P., Watson R. T., Kinney S. T., *Computer Augmented Teamwork - A Guided Tour*, VNR Computer Library.

COMIC (1993), Issues for Supporting Organisational Context in *CSCW Systems*, Bannon L., Schmidt K. (eds.), COMIC D 1.1, Esprit Project Research Project 6225.

DeSanctis G. L., Gallupe R. B. (1987) A Foundation for the Study of Group Decision Support Systems, *Management Science*, 33, pp. 589-609.

Dickson G. W., Poole M. S., DeCanctid G. (1992) An Overview of the GDSS Research Project and the SAMM System, in Bostrom R. P., Watson R. T., Kinney S. T., *Computer Augmented Teamwork - A Guided Tour*, VNR Computer Library.

Dickson G. W., Pertridge J. L., Robinson L. H. (1993) Exploring Modes of Facilitative Support for GDSS Technology, *MIS Quarterly*, pp. 173-195.

Egger E., Wagner I. (1992) TIME-Management - A Case for CSCW, in Turner J., Kraut R. (eds.), *Proceedings of the 1992 Conference on Computer Supported Co-operative Work*, Toronto.

Fisher A. (1981) *Small Group Decision Making*, 2nd Ed., McGrawHill.

Gärtner J. (1992) CATS-Computer Aided Time Scheduling - Ein Modell für die computerunterstützte (Schicht-) *Arbeitszeitplanung*, Dissertation, TU-Wien.

Gershon N. (1994) From Perception to Visualisation, in L. Rosenbaum et al (eds.) *Scientific Visualisation - Advances and Challenges*, London, Academic Press.

Heron J. (1989) *The Facilitators Handbook*, Kogan Page, London 1989.

Knauth P. (1987) Computergestützte Gestaltung diskontinuierlicher Schichtpläne nach arbeitswissenschaftlichen Kriterien; *AFA*, pp. 221 - 226.

March J.. G. (1991) How Decisions Happen in Organisations, *Human Computer Interaction*, pp. 95-117.

McGrath J. E., Hollingshead Andrea B. (1994) *Groups Interacting with Technology*, Sage.

Nachreiner F., Ling Q., Grzech H., Hedden I. (1993) Computer Aided Design of Shift Schedules, *Ergonomics*, pp. 77-83.

Olson G. M., Olson J. S., Carter M. R., Storrøsten M. (1992) Small Group Design Meetings: An Analysis of Collaboration, *Human Computer Interaction*, pp. 347-374.

Rabenstein R., Reichel R., Thanhoffer M. (1990) Das Methodenset - 5 Bücher für Referenten und Seminarleiter, *AGB*, (4. (ed.).

Schönfelder E. (1992) Entwicklung eines Verfahrens zur Bewertung von Schichtsystemen nach arbeitswissenschaftlichen Kriterien, Peter Lang Verlag.

Viller St. (1991) The Group Facilitator: A CSCW Perspective, in Bannon L. et al (eds.), *Proceedings of the Second European Conference on Computer Supported Co-operative Work*, Amsterdam.

Vogel D., Nunamaker J., Applegate L., Knsynski B. (1994) Group Decision Support Systems: Determinants of Success, in Gray Paul (ed.), *Decision Support and Executive Information Systems*, Prentice Hall, Englewood Cliffs, New Jersey.

Wolf W., Vollmann K. (1992) Arbeitszeit 1991: Regelarbeitszeit, Überstunden, Wochenend-, *Nacht- und Schichtarbeit*, Statistische Nachrichten.

Wood J., Silver D. (1989) *Joint Application Design*, John Wiley & Sons.